The Photon Storm Model

An alternate model of the universe that answers questions the classic model cannot.

Matter Creation   Gravity   Electromagnetism

The photon storm model validates Newton, Einstein, and quantum calculations using rational math and easy visualizations.

"The benefit of a simple model is that everyone can play with it."

By R. Bourne

Copyright 2017

To encourage a discussion: Imagine a universe that may or may not be our own. Imagine, for a moment, a universe filled completely by a storm of invisible photon energy moving forcefully in *all* directions, a photon storm universe.

THE CONCEPT IS NOT REVOLUTIONARY. Einstein coined the term, cosmic constant, to honor the possibility of such a field. Dark invisible energy has long been speculated. And most of the scientific community *now accepts* a Higgs energy field that must extend through the entire universe.

Physics would be much simpler in this photon-storm universe— no inexplicable force fields; and quantum theory, electromagnetism, gravity, and relativity would all be easily united. A unified field theory is easily created in any universe that possesses one thing, a unified field.

THE FAMOUS DOUBLE SLIT EXPERIMENT: The standard universe can only explain the double slit experiment with complex statistical formulas. An elemental particle must take all routes to its destination so that it can interfere with itself and thus create the standard interference pattern. But the same experiment is not at all complex in the photon-storm universe—the phenomenon is easily visualized and the math is simple.

TO UNDERSTAND the photon-storm concept merely requires we envision a universe filled completely with an intense energy field of stable photons too small to be easily detected. One tiny photon would possess near zero energy, but with a diameter approaching zero, this photon could travel side by side in massive numbers, and these numbers would add up to an extraordinary force. The photon-storm explains the *strong* force in the PS universe and gives it a number.

SPIN THEORY -- And pivotal to the photon storm concept is its logical consequence of spin interaction. In spin interaction, we find an excellent reason why the photon storm is contained in its own space and not dispersed through the void. Spin interaction also explains the meaning of light speed and the reasoning behind its affects including validating Einstein's mathematics and the weak force.

POINT CONSTRUCTION -- AND FINALLY, the photon storm model leads to an understanding of how elemental particles came into being, and how these particles combine--obeying simple rules to create a stable photon-storm universe which may or may not be our own. *Most important, the model explains precisely why particles later behave as they do—creating electromagnetism, radioactivity, and chemical reactions.*

The Photon Storm –
An intense storm of photons
moving in _every_ direction.

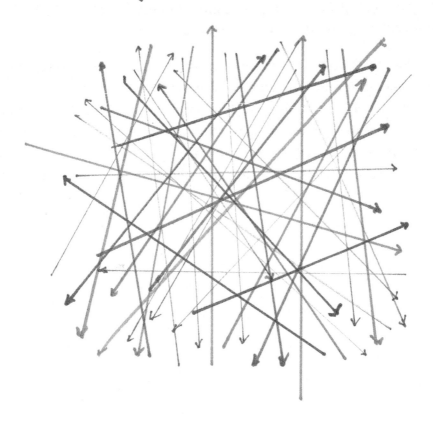

TABLE OF CONTENTS – THE PHOTON STORM MODEL

CHAPTER 1 -- TO ENCOURAGE A DISCUSSION (CONT):

In the Photon-Storm universe, space is not a vacuum. On the contrary, a nearly inconceivable amount of photon-storm energy fills this universe totally; and this photon energy moves forcefully in every direction, so it creates a significant affect everywhere equally, from deep space to the center of a planet.

As in our universe, photons in the Photon-Storm universe typically do not interfere with each other's movements. Also in common with our universe, the residents of the PS universe cannot easily detect photons of long wavelengths or exceedingly short wavelength; such photon particles can only be noticed when they interact with other particles that can be measured. So most photon-storm activity goes totally unnoticed.

Most of what we call *matter*, is actually quite empty. The space in one atom corresponds to the space in our entire solar system. So tiny photons can pass through entire planets without much interaction, but being thin and compact they can travel as closely together as they wish, in such numbers that they can pack quite a wallop when they actually meet something head-on.

The concept, then, is that these unseen, undetected photon-like spears exist in such numbers and so closely aligned that they can greatly affect the entire universe. This thread-like photon has also been named the God-photon, meaning no disrespect, but only that, *if it exists*, the God-photon is omnipresent with its great force felt equally in all directions everywhere and at all times.

Does this sound too far fetched?

TO GENERATE A BIT OF INTEREST, PLEASE CONSIDER THE PHOTON-STORM SOLUTION TO THE DOUBLE-SLIT EXPERIMENT, AN ENIGMA THAT CONTINUES TO PLAGUE ALL PHYSICS STUDENTS IN THE STANDARD UNIVERSE. HOPEFULLY THE READER HAS ALREADY ENCOUNTERED THE FAMOUS PROBLEM IN AN ELEMENTARY PHYSICS CLASS.

THE DOUBLE SLIT EXPERIMENT IN THE PS UNIVERSE

What is the double slit enigma? In 1804 a physician, Thomas Young, noted light waves formed a broad even scatter when sent through a single slit; but when the same light was sent through two closely aligned slits, the resulting light patterns interfered and formed lines of light and dark.

The conclusion was, since light interfered with itself in wave patterns, light must be a wave! Great!

Later however, when electrons or photons were sent singly through double slits, they interfered in the same way as if double waves were moving through the slits simultaneously. How could single electrons or photons create an interference pattern when only one wave/particle at a time was moving through a slit??? Can a wave know in advance that it will find interference in the future??

Einstein argued for some rational, unknown explanation, but other physicists, like Bohr, found adopting uncertainty mathematics—in which the particle effectively takes all routes between points and thus interferes with itself—could acceptably solve the problem. Physics students were required to ignore what could not be visualized. And so the situation in the standard universe remains.

Due to the quantum-wave duality, physics students are advised no models can be constructed of atomic particles—something that is both a wave and a particle cannot be visualized--only abstract mathematics can explain the standard universe.

In the PS universe, however, the double slit experiment is easily explained, since the mechanism involved is easily visualized.

In the PS universe, mass, gravity, and the movements of atomic particles are all determined by interactions with the energy of the Photon Storm. In particular, the movements of photons, electrons, and molecules are determined by their spins interacting with the storm.

Therefore, in the PS universe, the movements of photons and electrons passing through slits have little to do with the photons or electrons—*the results are pre-determined by the photon-storm and the slits!!*

(See following picture)

DOUBLE SLIT EXPERIMENT

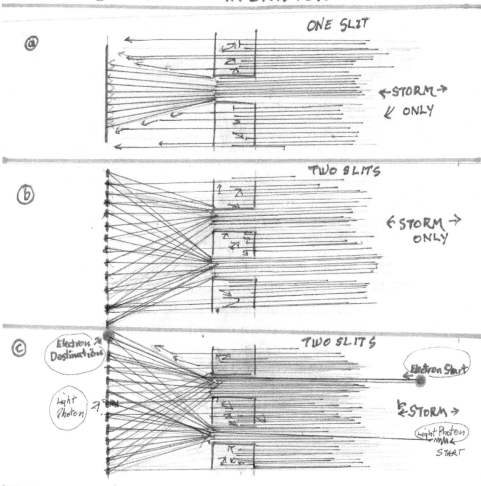

See in picture a, the photon storm passes through one slit. Since fewer photons are absorbed going through the slit than going through the barrier, the photons going through the slit are more dense, so they must spread slightly to even the density around them, making up for the photons reflected by the barrier. The picture only shows the activity from one direction, but it is equally true in all directions.

In picture b, the photon-storm passes through two slits, and like the classical slit experiment, the waves interfere with each other to create alternating slats. BUT THIS PATTERN IS TOTALLY INVISIBLE! OUR EYES DO NOT SEE THE PHOTON STORM IN THE SAME WAY WE SEE LIGHT! WE SEE NOTHING THERE.

Finally, in picture c, visible light photons or electrons are sent through the slits. These individual wave/particles are not interfering with themselves; on the contrary, the photon-storm is guiding them to align with the interference pattern already created by the storm.

In the PS universe, no enigma exists. If we recognize the power of the storm, the problem is easy to visualize. We do not require Feynman's quantum math and mentally impossible constructions to explain the phenomena.

In the Photon-Storm universe, Einstein would BE RIGHT!

And if the double slit experiment has a simple explanation easily seen, then many other phenomena in the PS universe—deemed too complex for the human mind in the standard universe--may also be easily visualized.

SO:

In the Photon-Storm universe, the language of exotic mathematics is less important than the ability to draw pictures on a kindergarten level.

To repeat: The PS universe is rational. All causes have effects, and though elemental particles may assume different forms, their most stable forms may be *visualized*.

The Photon Storm model will later demonstrate, through the spin concept, why the photon-storm is so important in the movement and construction of atomic particles.

The model will also show visually the importance of light speed in the PS universe, and more importantly, why light speed acts as it does.

We can also see precisely how the photon-storm creates mass, gravity, electromagnetism, and the weak and strong forces in the PS universe. And what do these forces actually look like?

Further, how did matter and the electron come about in the PS universe? How is a proton formed and why is it so amazingly stable? What is fluff? And how does it relate to the Wo, W-, and Zo photons. Please read further.

BUT IT IS NOW NECESSARY TO START AT THE BEGINNING: WHAT IS _MASS_ IN THE PHOTON STORM UNIVERSE?

CHAPTER 2- HOW DOES THE PHOTON STORM CREATE MASS?

The first effect caused by a photon-storm is *mass*. In the photon-storm universe, mass is <u>*not*</u> a given; mass is a <u>reaction.</u>

A STORM OF PHOTONS FROM ALL DIRECTIONS CONFINES A PARTICLE AND GIVES THE PARTICLE SHAPE AND INERTIA, OR MASS.

MASS is created *by the force of the photon-storm—also referred to as the cosmic constant/,Higgs,/black energy field--on all particles, creating inertia.*

To restate, in the PS universe, a particle does not possess mass simply because it is a particle; a particle only possesses mass because it is confined by the Photon Storm.

Some would argue, a photon cannot deliver an energy force to confine a particle, since the photon has no mass--but note, the force is being delivered to a particle that is initially equally massless.

In the Standard Theory in the standard universe, mass is a measurement of an object's inertia (resistance to movement). No explanation is given why an object <u>has </u>inertia other than that an object has mass, which may be circular reasoning?

Recently, the Higgs field has appeared as an explanation for how a particle receives mass through interaction in the standard universe, though the method is not clear.

On the contrary, the mass phenomenon in the photon-storm universe is easily seen—it is an interaction between a particle and the storm.

Because mass in the PS universe is an interaction, it is not a thing all particles possess. Any particle that, for any reason, does not interact with the photon-storm/Higgs-black energy field would have <u>no</u> mass. Equally, a huge planet outside the photon-storm universe, beyond gravity, in a true vacuum, would possess no mass. Not only would such an object have no mass, it soon would have no shape, dividing immediately into massless photons.

A storm of trillions of god-photons each moment on every particle is necessary to give any object its shape in the first place—at least, this is the case in the Photon-Storm universe we are imagining.

So, how would mass be calculated in the photon-storm world? WARNING: MATH ALERT. What forces are in play here?

MASS MUST BE A FUNCTION OF:

S = THE FORCE OF THE PHOTON-STORM/PER CUBIC AREA OF A PARTICLE

Rf = The particular Reflectivity/absorption of that particular particle.

A3 = The area cubed of the particular particle.

$$MASS = (S)(Rf)(A3)$$

A Particle bounces, absorbs, and
emits photon particles by the
trillions, creating mass and inertia!

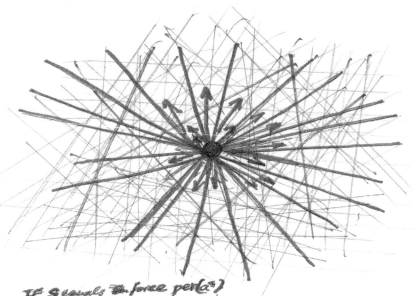

If S equals the force per(a^3)
(S) (area) (R) = Mass of object

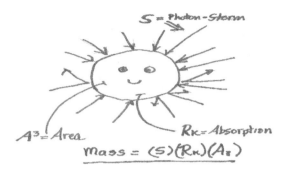

$S = \text{Photon-Storm}$

$A^3 = \text{Area}$ $R_K = \text{Absorption}$

$$\text{Mass} = (S)(R_K)(A_3)$$

A particle bounces, absorbs, and emits God-photon particles by the trillions each instant, creating mass and inertia.

MASS = (FORCE OF STORM) (REFLECTION) (AREA)

MASS = (S)(Rf)(A3)

<u>**So, to repeat, mass is not innate: A particle receives its mass from its interaction with the photon-storm, at least in the photon-storm universe.**</u>

Values are not now available for (S) or (Rf), but they will appear in a few pages. (S), the value of the storm is a very, very large value. (Rf) or reflectivity is a tiny, tiny percentage in normal objects, such as planets and human beings. However, at the atomic level or in black holes, the reflectivity can change radically, reaching to 100%. And (Ra) or absorbance, in which energy is actually absorbed by particles, can also change at atomic levels.

But for *normal* objects of known area, mass is simply equal to the photon-storm hitting it from every direction each instant times normal reflectivity.

MASS = (FORCE OF STORM/AREA)(AVG REFLECTIVITY)(AREA3)

SO WE KNOW WHAT MASS IS IN THE PS UNIVERSE, WHAT IS GRAVITY?

CHAPTER 3 – HOW DOES THE PHOTON STORM CREATE GRAVITY?

In the photon-Storm universe, gravity is created by a different method than the standard model utilized by Newton or even the revised model described by Einstein.

Newton's model contends that all matter sends out gravity waves that attract all other matter. These waves cause matter to attract and accelerate to each other according to the formula: $F = (m1)(m2)G/d2$. The masses of the two objects times G, which is the gravity constant, divided by d2, which is the square of the distance between the two objects.

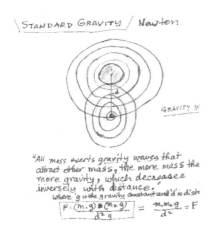

STANDARD GRAVITY / Newton

GRAVITY W

"All mass exerts gravity waves that attract other mass, the more mass the more gravity, which decreases inversely with distance."
where 'g is the gravity constant and 'd' is dist

$$F = \frac{(m_1 \cdot g)(m_2 \cdot g)}{d^2 \cdot g} = \frac{m_1 \cdot m_2 \cdot g}{d^2} = F$$

Einstein altered Newton's model to encompass a new idea of gravity, that of a time-space continuum. Gravity is created because mass distorts time-space, bending space-time so that straight lines curve inward, causing objects of mass to accelerate to each other. Einstein's formulas condense to the same as Newton's when objects are of reasonable size and traveling at reasonable speeds, like most of the objects in our solar system. Only around huge objects, like stars and black holes, do Einstein's formulas make a difference. Only at great masses are the space-time lines distorted enough to make Newton's calculations inaccurate.

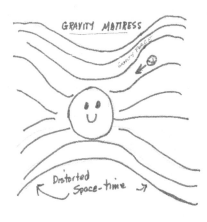

Einstein's calculations treat space-time as if it were a mattress, and when something heavy bends the springs, other objects are drawn its direction.

GRAVITY IN THE PHOTON-STORM UNIVERSE: Stuff is easier to visualize here.

Gravity in the photon-storm universe is caused by a completely different mechanism--when any two objects shield each other from the photon-storm, they are pushed towards each other along the shadow between them.

GRAVITY

TWO OBJECTS SHIELD EACH OTHER
AND ARE PUSHED TOGETHER BY PHOTON STORM

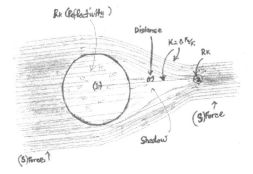

Although the photons come from every direction, the drawing above, for simplicity, shows only the flow from opposite directions. As if the photons are winds blowing from opposite directions, the objects are blown toward each other through the shadow they create by shielding each other from the wind's full force.

All objects with mass absorb or bounce part of the photon-storm, and so all objects with mass acquire gravity.

HOW DO WE COMPUTE GRAVITY IN A PHOTON-STORM
UNIVERSE? ACTUALLY, IT IS QUITE SIMPLE. *MATH WARNING HERE.*

Photon-storm gravity is a function of:
       (S), the force of the photon storm/per area.
       (Rk) the reflection/absorption of the particle's involved
       ^K, the difference between the force lines within and without the
shadow, or Ko/Ki [This is a variable of how quickly the photon-storm
realigns.] This variable is also the same as ^%/d, the percentage change
of the (S) field as it realigns from a shadow state to a full (S) stage over
distance.
       (A3) the cubic areas of the objects involved.
       (r) the distance between the two objects.

We already utilized S, Rk, and A3 in computing mass.
       For one object the formula is (S)(Rk)(A3)(^K/d)/r2= Force of
acceleration.
       For two objects, a and b, the formula is: Force of attraction=
[(S) (Rk)[(A.)3](^K/d)][(S)(Rk)(A2)3 (^k/d)] divided by r2.
       But in the Photon-storm universe (S)(Rk)(A3)=Mass, so the
formula condenses to;
       (Ma)(Mb)(^K/d)2/r2

       But the above formula is exactly like Newton's
       (M1)(M2)(g)/r2=Force

Therefore, assuming (^K/d)2=G, then the formulas are identical. Gravity acts in the photon-storm universe precisely as it does in our standard universe.

Using the earth's mass and cubic area, and giving (Rf)
the guesstimate value of 4(10)-13. This is roughly the
amount of empty space in a single atom. Using the
earth's mass and cubic area, we can solve *roughly* for

(S) = $[1.493]x(10)28th$ **kg/cu km. This is a *very* big number. Why people in the PS universe are not crushed to nothingness is because their reflectivity is quite small and the (S) force is exerted in all directions and only on elemental particles in their bodies.**

GRAVITY,

TWO OBJECTS SHIELD EACH OTHER
AND ARE PUSHED TOGETHER BY PHOTON STORM

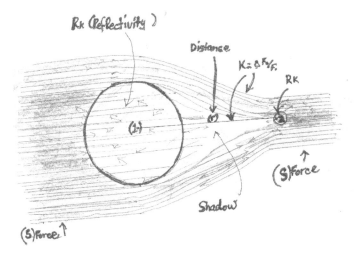

Rk (Reflectivity)

Distance

$K = \Delta^{k}/F_{s}$

RK

(1-)

(S)Force

Shadow

(S)Force

HOWEVER, AT HIGH SPEEDS AND EXCESSIVE GRAVITIES, SUCH AS AROUND LARGE STARS AND BLACK HOLES, THINGS MUST CHANGE.

Rf must change because large stars and black holes have a density that reflects and absorbs more than the normal amount of the photon-storm.

^K also changes around large stars or black holes, since the distortion of the photon routes are much greater, and the shadow effect is more extreme.

The changes above are the same changes Einstein uses to correct Newton's equations when they are near heavy gravities.

SO:

WHETHER SPACE-TIME IS A VACUUM WARPED BY MASS TO CREATE GRAVITY, OR WHETHER SPACE IS A THICK ENERGY FIELD THAT DEFINES MATTER AND IN THE PROCESS WARPS TO CREATE GRAVITY, *THE RESULTS ARE THE SAME*!!

Repeat:

WHETHER SPACE-TIME IS A VACUUM WARPED BY MASS TO CREATE GRAVITY, OR WHETHER SPACE IS A THICK ENERGY FIELD THAT DEFINES MATTER AND IN THE PROCESS WARPS TO CREATE GRAVITY, *THE RESULTS ARE THE SAME*!!

Could the similarities between the PS universe and the standard universe <u>all</u> be coincidence? Or does the evidence suggest the need for additional scrutiny? Surely, continuing the discussion can harm nothing. And the process should be stimulating.

CHAPTER 4: SOME IMPORTANT FORCES IN THE PS UNIVERSE—Entanglement and Spin

Gravity is a limited force. The power of the photon-storm is great, but gravity only reflects a tiny, tiny portion of that power. Only the reflected part of the storm becomes part of a particle's gravity impact in the PS model.

Most of the storm passes completely through large particles, like planets, with little consequence. However, besides being reflected, part of the storm may be absorbed within atomic particles; this energy must later be emitted as similar energy. Such absorbed and emitted energy would have an affect on an object's creation of a gravity shadow. At a stability point, the amount of energy absorbed would be cancelled out by the energy emitted, assuming the energy is emitted in the same random fashion as the energy is absorbed.

Entangled energy might add to the inward pressure the storm exerts on atomic particles and departing energy might have the same effect.

This concept is based on the assumption that the Reflectivity (Rf) of a particle is not directly based on its absorption (Ra), that the two phenomena are separate. If this assumption is correct, then mass, gravity, electromagnetism, and the strong force are all directed by the photon-storm force, and would be united in the PS universe.

Finally, if reflectivity is not always aligned with density, then measured mass could be hidden in particle bundles. Such photon-particle bundles could assist in the process of keeping atoms together and would appear fleetingly during particle destruction or construction.

We will later discover an even stronger force the storm uses when putting elemental particles together, directionality. This concept requires more explanation and will be demonstrated later.

ENTANGLEMENT:

An additional force holding an atomic particle together in the PS universe would be the *possibility* of photon entanglements, meaning the combinations of mutual spins. If spins can combine and twist together, they could exert an additional force to keep atomic particles together. These points could also entangle photon-fluff. We will get into a lot of fluff later.

POINT ENTAGLEMENT:

And if two or more photons meet at the exact same point, they could be united strongly, very strongly by the PS storm. Such a point particle will be discussed later. (Ch. 9. How is matter created? Pg. 62) Point entanglement explains why quarks and protons are stable.

For atomic particles, mass roughly equals the storm force (S) times the total reflectivity, or M = (S) (Rf) (Ra).

And the total force holding a particle together is equal to the storm force times the absorbance, times the reflectivity, times the entanglement force, or Ft=(S)(Rf)(Ra)(En).

If absorbance were a percentage, maximum inward force would occur at 100%.

If spin entanglement were neutral at 1, any value above 1 would increase the total force; any value below one would decrease the total force.

Black holes belong to a further class of matter absorbing more energy than they are capable of emitting. However, according to the PS model, black holes would emit energy at their surfaces like other particles. *If the spin concept is correct, outward movement of energy would be facilitated by energy moving inward. (Spin concept, read further)*

Later, the configurations of atoms in the PS universe will reveal a much simpler manner to compute atomic masses, and answer the question of why quarks, electrons and protons weigh what they do. (Pg 98)

SPIN: A WEAK FORCE???

Also of note, forces pressing *outwards* in an atomic particle must equal all the forces pushing in. Spin/spring energy is a force that must constantly push outwards struggling to unravel all atomic particles. Complex atoms with multiple spin systems rotating simultaneously may spin into unstable structures at random moments. Such unsteady configurations, when they occur, could cause atoms to spin apart.

Supposing such an unstable spin configuration occurred randomly at ten-year intervals, the complex atom would have a half-life of ten years.

Assuming spin energy is conserved, such an outward pressing force would keep a black hole from assuming a zero area.

And if a black hole were to lose the (S) force pushing it together, and it exploded outwards, the weak force would be the power behind the explosion. In this case the weak force would be anything but weak.

In fact, in the PS model, the weak and strong forces end up being different configurations of the exact same energy. Keep reading.

Reflected photons create
mass and gravity.
Absorbed/emitted photons help
keep particle together.

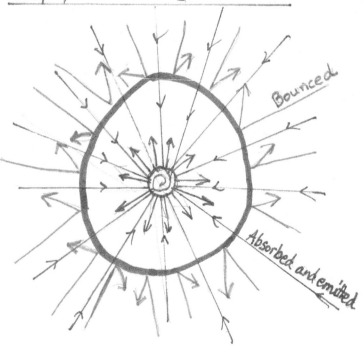

CHAPTER 5 -- CHANGES IN THE MODEL AT VELOCITY AND AT ATOMIC LEVELS IN THE PHOTON-STORM UNIVERSE – VALIDATING EINSTEIN

PHOTON STORM GRAVITY: ANY TWO OBJECTS THAT SHIELD EACH OTHER FROM THE PHOTON STREAM WILL BE PUSHED TO EACH OTHER ALONG THE PHOTON SHADOW BETWEEN THEM.

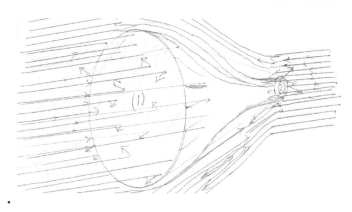

Although the people will still be getting wet from photons moving sideways, they will be getting a little less wet from the areas protected by reflective umbrellas. And this small force will push them together.

-

• <u>SOME CHANGES AT VELOCITY AND AT ATOMIC LEVEL</u>

- $K = {}^\wedge Fo/{}^\wedge Fi$ As Fo approaches Fi, K approaches 100%, since almost all photons are absorbed. Gravity increases drastically to the limits of the S force.
- Ra, Rf absorption reflection in black holes or at high speeds also increases, perhaps approaching 100% and forcing particles ever smaller.
- Causing mass/gravity to increase exponentially per area
- As speed increases, opposing S also increases against the direction of accelleration and decreases in the opposite direction since speed creates its own shadow.
- ${}^\wedge$Sfront/${}^\wedge$Sback. ${}^\wedge$Sfront near S, ${}^\wedge$Sback nears 0.
- Causing height to flatten and width to widen and physical processes to slow approaching the speed of light, depending on the presence or absence of acceleration.
- Sd or Storm-density increases with velocity. $Sd = V/C$

<u>TIME CHANGES</u>

Time slows as a rocket in the Photon-shower universe approaches light speed. Imagine a photon clock in the rocket, but as the velocity of the rocket increases, the distance traveled of the photon also increases. Since the speed of light is constant, in or out of the rocket, time must slow down compared to an outside clock with a shorter distance.

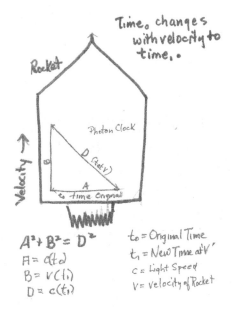

Time₀ changes
with velocity to
time₁ •

Rocket

Photon Clock

D (tatv)

Velocity →

A

t₀ time Original

$A^2 + B^2 = D^2$
$A = c(t_0)$
$B = v(t_1)$
$D = c(t_1)$

t_0 = Original Time
t_1 = New Time at v'
c = Light Speed
v = Velocity of Rocket

HAPPY MATH AGAIN

For those interested in the math, here is some. Otherwise just turn the page.

To compare the original time, To, to the time in the rocket, T1, we need to compare the length of the original clock, line A in the drawing, to the new clock length, line D in the drawing. Since the lines are connected by a right angle triangle. A2 + B2 = D2. The speed of light is (c); velocity is (v). Velocity times time = Distance.

$$A2 + B2 = D2$$
$$(cTo)2 + (vT1)2 = (cT1)2$$
$$c2To2 + v2T12 = c2t12$$
$$(To)2 = c2(T1)2 - v2(T1)2/c2$$

$$(To)2 = (T1)2 - \left(\frac{v2}{c2}\right)(T1)2$$

$$(To)2 = (T1)2 \left(1 - \frac{v2}{c2}\right)$$

$$(To)2/\left(1 - \frac{v2}{c2}\right) = \quad ((T1)2$$

$$\frac{(To)}{\sqrt{1 - \frac{v2}{c2}}} = (T1)$$

As v approaches c, the speed of light, T1, time in the rocket, approaches infinity.

Height varies depending on whether an object is being accelerated. A rocket accelerating will be wider and shorter like a compressed spring, but once the acceleration ends, the object will spring back in the opposite direction becoming long and thin, as if it were accelerating into a black hole. Height Original is to Height at velocity. Hto=Original Htv=at velocity.

Htv = Hto(c+v)2/(c-v)2 (K)

$$Htv = Hto \frac{(c + v)2}{(c - v)2} (K)$$

Assuming the particles in the rocket will maintain their same area3, then Length1 times Length 2 times Height stays constant. The ratio of L1/L2 would also stay the same. (L1v)(L2v)(Htv)=A3

$$(L1v)(L2v) = \frac{A3}{Htv} = A3/(Hto)(c + v)2/(c - v)2$$

So, knowing original A3 and the L1/L2 ratio, the new lengths at velocity can be calculated, and they would become very small as (c-v) approaches zero and Height approaches infinity.

So the time changes at near light speed in the photon-storm universe would be exactly equivalent to those changes in our own universe according to Einstein.

Is the similarity again a coincidence? Or should a greater kinship be suspected?

PHOTON STORM FRICTION

In the standard universe, empty space is frictionless because it is empty. In the PS universe, however, space is not empty. This friction increases around massive objects that condense storm properties in the same way storm properties increase as the velocity of an object approaches to near light speeds.

PS friction has the same affect as squeezing time-space. As density increases, either of time-space or the storm, time slows. Viewed another way, a planet is moved by two vectors, one of velocity sideways, and the other, a gravity push towards the object around which the planet orbits. These vectors change little in a circular orbit, but they change quite a bit in an oval orbit that approaches very closely to a massive object. The velocity vector of a particle decreases slightly as it enters a densely packed storm area, but receives a slight boost as it exits that area. The gravity vector increases acceleration as a planet nears its massive object and decelerates as it moves away.

These vector changes may explain perceived orbit anomalies noticed, for example, in the planet Mercury. The change is the same as moving in and out of an area of warped time-space.

CHAPTER 6– MORE SPECULATION
THE END OF THE PHOTON-STORM UNIVERSE

Speculating about the end of the Photon-Storm universe may seem premature, since we have only recently imagined it. But if it were to exist, its end would be rather predictable. In its middle would have been created a great black-hole conglomeration, a very condensed object that absorbs all the energy the photon-storm sends at it. *In fact, this object after billions of years finally absorbs the photon-storm completely.*

But in the Photon-Storm universe, gravity and mass are <u>creations</u> of the photon storm, not aspects simply of matter. So this great black hole conglomeration suddenly finds itself without gravity, and even, without mass.

Repeat: This tiny spec that contains the entire universe would have no mass or gravity.

But soon, all the energy the dark object has patiently absorbed over the millennia's suddenly becomes unstable. Nothing except conjoined spin is holding this energy in or keeping it from doing whatever it desires--but this spin can quickly unwind and does. The resulting expansion would surely be very large.

Questions remain: Will the new universe unfold exactly as it folded? Or will small permutations in the fabric of creation, create something much different?

Interesting to speculate, if we were part of an ancient Photon-Storm universe, might we be recreated a trillion years hence; and if so, might we even be a bit improved?

We would hope so.

CHAPTER 7--FROM THE BEGINNING
THE ELEMENTAL PHOTON – Why it moves!

In the standard universe, photons move through the vacuum of the universe at the speed of light. We don't know why photons move in the standard universe; we only know they move.

The PS universe, however, is rational. If photons move, there is a reason, and this reason *must* be fairly simple—the PS universe is a simple universe. Why do photons in the PS universe move forward at the speed of light? Well, they push each other.

This explanation may seem absolutely ridiculous, but in the PS universe, a vacuum does not exist. The PS universe is filled with photons; therefore photons have something against which they can *always* push—each other.

How does this push come about? Well, that depends on the specific design of the photon. The simplest photon design is simply a wave. Assume a simple photon is a wire energized by a frequency that moves at light speed one direction from its pointed head to its tail—also assuming a photon possesses a head and a tail. (See pic., next page.)

Note in the first picture a single wave photon is pushed by its wave movement forward. The key concept is that this PS photon is actually physically pushing itself forward by pushing backwards against other photons.

In the second picture, (b), two photons move past each other in the opposite direction. But kindly, as they brush by each other, they give each other something off which to push, each pushing the other in the opposite direction. So in the PS universe photon interactions actually assist forward movement.

Photon

Direction

Push

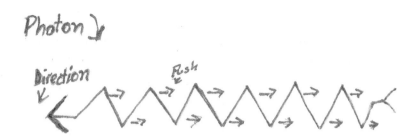

a)

b)

Direction

Opposite Push

Direction

2 Photons Passing

This concept of communal movement is extremely important in the PS universe. Imagine several surfers paddling in the opposite direction. If their hands meet pushing opposite directions, they would each help push the other forward in the opposite direction. Such assistance would be particularly helpful if the surfers were in space and had nothing but each other to push off of.

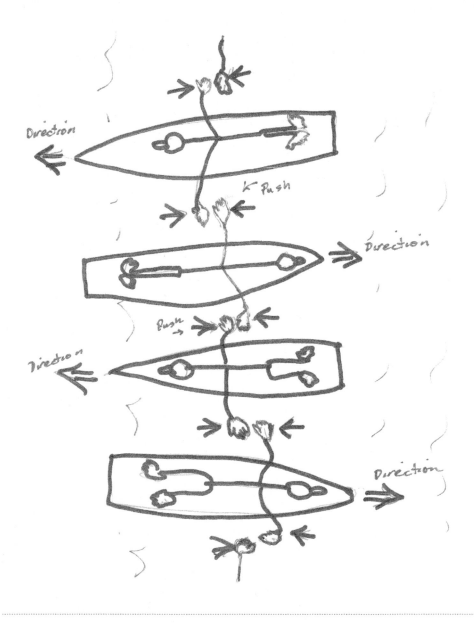

So, the Photon Storm universe is a helpful universe. Not only surfers, but even photons assist each other as well as they can. This concept, of course, assumes the photons in the PS universe are actually particles as well as waves and have a substance that can actually push. Thankfully, they don't need to push very hard. Fellow photons, after all, have no mass, and zero friction forward.

If this concept—call it the shared spin concept—is accepted, then it explains a great deal of phenomena in the PS universe, particularly the effect the photon storm has on fellow photons and why some photons choose to move or twist together. These concepts will reveal themselves in more detail later.

We apologize for the simplistic drawings. As physicists in the classic universe advise us, the photon quantum form cannot fit easily into our mind's eye; flitting between a particle and a wave, it cannot be visualized. According to most authorities only blithering idiots would seriously try to draw a photon. We may be idiots, but we do not blither.

Fortunately, we are imagining a photon-storm universe, and in this universe concepts are simpler. The PS universe is entirely mechanical and rational. We are free to imagine energy and the photon as tangible items. And though we are advised the photon and all elemental particles have numerous possible quantum states of being, we enjoy the knowledge that in any system of random possibilities, the most stable state will dominate, according to the equation:

Q Probability: $Q = Q_1/t_1, Q_2/t_2, Q_3/t_3 \ldots$

The possibility of Q will depend on the most stable state over time.

To explain, if an elemental particle can exist in, say, 5 states, but the first four states only last for microseconds, and the fifth state lasts for a minute, if we examine a system involving the particle Q after a few seconds, we will find the system dominated by the most stable state, the minute state--and this state will increasingly dominate over time. This rule will prevail even in more complicated systems with many random configurations--assuming stable states exist.

So if we seek to know a particle with many random states, the most stable state through time is the one on which to focus. And if we can imagine it, we can draw it—if we have enough crayons.

SO START FROM THE BEGINNING, THE REAL BEGINNING

SEVERAL PHOTON DESIGNS ALL LEAD TO A SIMILAR PHOTON.

But what happened in the PS universe before the big bang? Let us speculate from the real beginning, when the Photon-Storm universe was energy and nothing else. Was this energy in the form of photon waves, or did something come even before photons?

THE ENERGY BALL POSSIBILITY

Putting the stability concept to work, imagine the universe when it was only energy. This energy, whatever it was—strings, waves, thread, bubbles--could exist in multiple exotic states. But its most stable state was probably that of a ball, a ball of energy.

Why energy? Well, without energy, there would be nothing.

Why a ball? Well, the sphere is the dominant form in our standard universe. Squirt water or molten metal in outer space, it will assume the form of a perfect sphere. Most feel this sphere creation is the result of surface tension; but the photon storm might also cause the sphere-situation. So, in the Photon-storm universe, the sphere is certainly dominant. God-photons hitting any particle from all directions will always have a tendency to turn particles into spheres.

OPEN LOOP OR CLOSED LOOP

In string theory, the closed or open loop is argued, but in the PS universe, the two designs end in the same configuration—a simple wave.

CLOSED LOOP DESIGNS

We aren't forgetting that energy could take many other forms beside spheres—it is free energy in a free democratic universe. It may have been in a stable waveform from the beginning. But its most stable earliest state *by intuition* would be an energy ball or loop, very tiny, but reasonably stable, and a reasonable state in which to begin our PS universe.

So these balls of energy—may have occupied a transitional state, for a few seconds—in the early Photon-Storm universe. They were the simplest and most stable units imaginable., even before waves. They filled up all space. And they all shared their energy with each other by bouncing continually against each other, spinning or vibrating at the same communal speed or tone--which turns out to be the speed of light--while simultaneously forcing each other into roundness.

DIRECTIONALITY

The only problem with balls of energy bouncing endlessly off one another would be that none of these balls could get anywhere properly. *They had plenty of energy, but no directionality.*

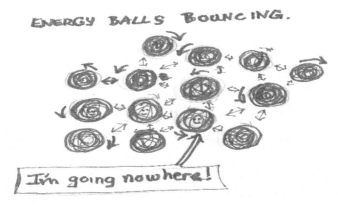

ENERGY BALLS BOUNCING.

I'm going nowhere!

To achieve directionality, the energy balls—possibly strings, or bubbles—needed to find a new stability.

POSSIBLE CLOSE-LOOPED ENERGY BALLS COMBINING
Interior return spin is shielded by exterior spin.

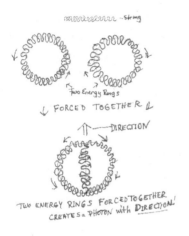

THE SPIN THEORY AGAIN: Simply enough, if two energy balls join together with outward spins moving the same direction, this outward spin will act exactly like oars on a boat, spinning against fellow elementary objects, pushing the combined energy balls forward in a single direction, and proceeding that direction at the speed of light. Return interior spin is shielded.

So, the model above would also act, once it is pressed into a long spear, exactly as a photon wave. In the standard universe, photons move through a vacuum at the speed of light. Why photons choose to travel at light speed is a mystery in the classic universe.

In the photon storm universe, photons move through a sea of fellow photons at the speed of light. The reason they travel at light speed is obvious. They are conjoined by their mutual exterior spins or wave front. Any spin or wave faster or slower than light speed encounters resistance from the storm. One standard speed spread through shared wave movement is the only choice.

The photon form must assume a long straight shape as it flies effortlessly and without friction through the universe. Each meeting with an energy ball assists it forward in the same direction it is going. And the same would be said if it meets other combinations like itself. If a backwards spin is exchanged, both objects are assisted to continue their routes forward and at the same speed as their mutual spins, the speed/tone of light.

And to further facilitate their movement forward they adopt clockwise or counterclockwise spins; such spins allow them to conveniently move over and under each other as they pass each other with the minimum amount of disturbance. Polarized spin will become even more important later.

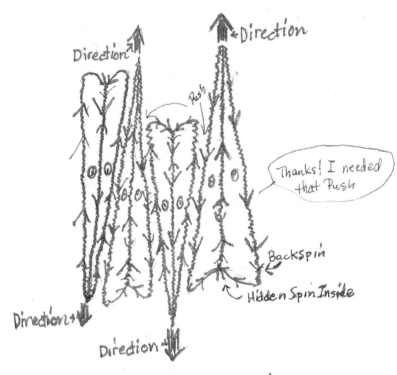

Photons Push Each Other
In the Opposite Direction

THE VIOLIN PHOTON – ANOTHER CLOSED LOOP DESIGN

A different photon design could result from a single violin-string energy ball, twisted into a figure eight, and then pressing forward. The resulting possible-photon would possess the same attributes as two combined energy balls—backward spin on the outside acting as oars to push it forward with forward spin buried internally.

(See picture next)

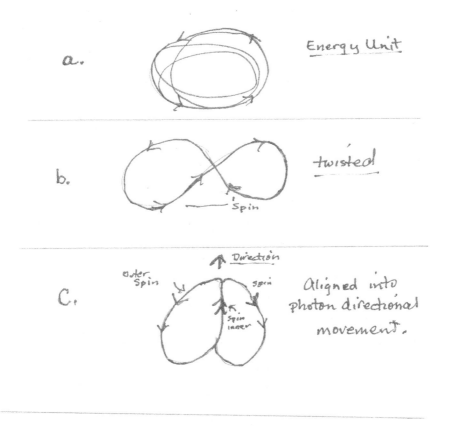

a.   Energy Unit

b.   twisted

Spin

c.   Direction   Aligned into photon directional movement.

outer Spin   spin   Spin inner

Backwards spin functions like oars to push photon forward through sea of fellow photons. Interior spin is buried.

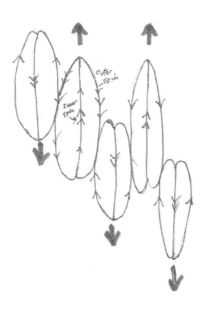

The picture above shows porpoise-like photon-like objects—all right, surf boards--pushing each other in opposite directions at the speed of light. ARE THEY

PHOTONS? In the Photon-Storm universe we will call them elemental photons, but they likely become very long and thin, pushed into long, long spears with near-zero width and no mass, polarized—still with their backwards spin to push each other forward—and slicing deftly through the universe like arrows.

**Coincidentally, if we accept the spin theory for the speed of light, we now SEE exactly why light speed is constant through the universe and has nothing to do with moving objects. The speed of light is totally set by the spin or wave movement of other photons and has nothing to do with the movement of objects from which light is received or projected—at least this is so** **in the Photon Storm universe.**

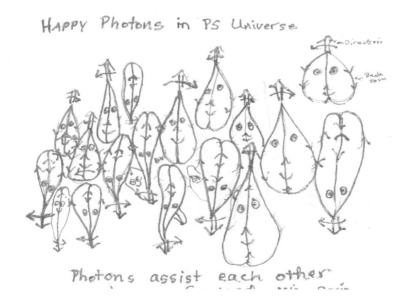

Happy Photons in PS Universe.

Photons assist each other

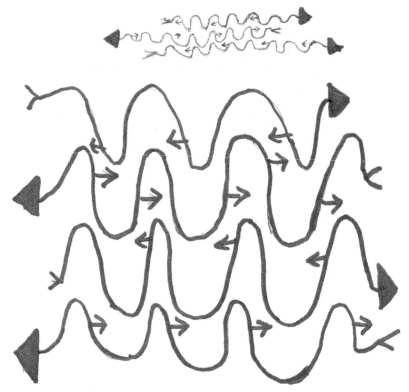

Photons push each other
in _opposite_ Directions.

OPEN LOOP DESIGNS

The previous picture is of a violin string OPEN LOOP or WAVE DESIGN. This is the simplest photon design and evolves from a single tight string with front to back wave motion. This motion acts in the same way as other PS photons pushing the photon forward through the exchange of a backward push. THE VIOLIN STRING MAY BE THE MOST LIKELY DESIGN.

The point is, each photon design leads to the same photon universe, so it doesn't matter which one of them is totally accurate. Several of them would work well, pure wave, combined energy ball, and twisted violin. We have our favorite—the helix design, pg 56, 57--but they all lead to the same destination. Anyone reading this should feel free to pick a favorite, or conceive his or her own design. In an indeterminate universe, they all may be correct.

A REVIEW...

Let us review for just a moment. We started with some elemental particles, possibly strings, balls, or simple waves; some of these elemental particles were together somewhere, side by side in a tiny universe. God came along and, wanting some music, banged this universe with a drumstick. The resulting sound was a vibration exactly at the speed of light. It sounded quite wonderful.

All the particles in this universe were now vibrating together wildly. Their vibrations began to push against each other. If the energy ball scenario occurred, the spherical form came to dominate.

Later these elemental spinning/vibrating balls—or simple waves-- combined or twisted into *even more stable* shapes, wave shapes

or combinations with outward and inward spin in opposite directions—
*this spin allowed our elemental particles or combinations to streak
forward at the speed of light and call themselves photons.*

We had a universe of photons, all pushing against each other,
happily moving shoulder-to-shoulder, and even happier when they
adapted sidespin polarity side to side to nicely pass over and under
each other in wonderful synchrony.

And finally, when some of these photons grew too
energetic, their interactions, reflecting and absorbing the storm of
fellow photons, forced them to attain states of mass, states of inertia
that limited their movement—these mega-photons attained mass, and
then their reflection of the storm created gravity. The rest is cosmo-
history.

But we need to slow down here. How exactly did tiny balls of
wave energy *combine* to create an entire PS universe? Or if wave
quanta began it all, how could they intensify to turn into particles?

The creation could have occurred in many ways, but in the PS
universe we feel it occurred in the simplest manner possible. As
Einstein inferred, *"God did not go out of her way to create enigmas in the
PS universe."*

But in the PS universe, we are at a disadvantage moving forward.
Unlike in our standard universe, we cannot have strange forces or
exotic particles—they don't exist there! We have only *one* thing! We
have the photon-storm and we have our newly engineered photons.
That's two things; but never mind. In order to create any possible
universe we are limited to these two things.

To repeat, the Photon Storm Universe is filled with tiny photons
all spinning and moving past each other without mass or friction at light
speed—nothing else, nothing—no exotic particles or strange force
fields, nada.

How can an entire universe evolve from so little?

A small digression: "Is this all there is?" A depressing jazz song was once popular by that title. But the point is: *Existence is not limited by what we find; it is only limited by what we can create with what God has given.*" And so the challenge is to utilize the simplest of energy forces, the photon storm and the god-photon, to create a complete, workable universe.

Many, many alternatives are available to choose from—perhaps too many. In fact, too many choices are definitely the problem. The reader must be the one to decide what is right or wrong. We will throw out some ideas. No assurance can be given of absolute truth.

Again, *in the PS universe*, the picture is simplicity and stability! We are depending on God to create a reasonable universe, and we are depending on the reader to bring us clarity when we go off course. So sharpen your crayons, and we can all help create an accurate picture! Thanks.

Here is a start.

CHAPTER 8 -- THE GOD-PHOTON INCREASING ENERGY

WHAT IS IT? High energy, low energy? Single violin string or mega-combination?

Clearly, if the god-photon exists, it must be extraordinarily elusive. Otherwise the facts about it would be well established to scientists in the PS universe years ago. How can it be so elusive? Well, it may be a photon with an incredibly long wavelength. Such is the main premise. Such a low energy photon would be nearly invisible since it would pass through almost everything without affect.

However, a high-energy photon with a very small wavelength might also be effective. Why this photon cannot be discovered might be because its wavelength is so very small, even less than Planck's constant, that current instruments do not detect it. Is this possible? If so, such a high-energy photon could easily keep the entire PS universe in line, create mass and gravity, and have plenty of power to re-combine in ways that would simulate the common forms of matter, quarks and so forth.

If such a high-energy particle existed in the PS universe, it could easily create a universe much like our standard universe. The only question would be, where is it?

However, without eliminating the high-energy alternative, the elusive low energy photon is also a simple way to build a universe. With a low energy photon we can build from the bottom up. And God may prefer simplicity. But this low-energy photon must exist in extraordinary numbers to create the effects necessary. In fact this photon would need to fill up the universe completely. It would only interact with objects when it had no other choice.

Fortunately, we live in the standard universe. And in that universe, brilliant scientists have uncovered many truths. We are fortunate to be able to lean on their efforts and assume some similarities to the PS universe.

Frequency times Planck's constant = energy

The frequency of a photon times Planck's constant equals the total energy of that photon. Or the speed of light divided by the wavelength of a photon times Planck's constant equals the total energy of that photon.

[(c)/wavelength)](Planck's Constant) = Energy

We can assume the same calculations rule in the PS universe.

WHAT DOES A GROWING PHOTON LOOK LIKE IN THE PS UNIVERSE??

REMEMBER: In the PS universe, we can visualize things, even photons that are massless and indeterminate.

WHAT ARE SOME POSSIBILITIES?

SINGLE VIOLIN STRINGS:

If a photon is a long rope, or a tight violin string, then it may simply increase its frequency to gain energy—still with backward spin pushing it forward at the speed of light. Such a design cannot be discounted and would have photons function in the PS universe very much like they function in the standard universe.

This design has many advantages; in particular, it is very simple. Imagine a long tight violin string resonating at ever increasing frequencies. Except for frequency differences and the associated wavelength changes of a photon, all photons would be the same. Packing these photons together closely might account for the extraordinary tension they would need to achieve high energy levels.

High-energy photons might encounter increased difficulty passing through the storm. They might begin to bounce low energy photons they encounter or even entrap them in their frequency folds, making a spinning mess that might one day acquire mass by its interaction with the storm. Eventually this high-energy photon becomes a different type of particle, a pre-electron or neutrino.

(See picture next page)

Note also, this violin electron has a configuration spin of ½, exactly like the configuration spins of electrons in the standard universe.

The Accordian Photon
(Violin Model)

Rotation

God-photons (long and thin)

High Energy Photon

Storm Interactions

Accordian Photon Very high energy

Storm photons

Accordian Pushed by storm to create mass.

The *single violin-string model* works quite well to create a universe. The only thing needed is to change the modulation and frequency of any single string to create nearly everything from the one elemental design. Easy.

A problem however is the conservation of energy. Where did the energy come from to originally change the violin strings from one energy level to another? If all energy was originally contained in the strings in an original PS universe, where did more energy originate to raise their energy levels; or where did the energy go if those energy levels were reduced?

Did God need to bang the gong again? Or was energy exchanged with a neighboring universe or somehow bartered within the universe to make the changes necessary.

So the *violin-string design* remains plausible, but the *combination design* is more affective in eliminating the need for foreign energy. Perhaps the state department should take notice.

PHOTON COMBINATIONS

In the Photon-Storm universe, thin, long low-energy photons may be twisted together in combinations. Such combinations would explain the *Planck quanta effect*, namely that photon energy does not increase linearly—as it would if a photon were strictly a wave function--but the energy increases in steps. Such step increases could easily be explained if each frequency increase were actually the addition of another elemental photon to a photon combination. The difference, then, between high and low energy photons would simply be the number (#) of god-photons in a bundle.

So—NOT EXCLUDING THE VIOLIN POSSIBILITY--high-energy photons in the Photon-Storm universe could actually be combinations of hundreds or thousands of elemental God-photons.

E=(#) (Planck's Constant)

Measurements of wavelength and frequency in the standard universe would then be measurements of a twisting combination photon helix in the PS universe

POSSIBLE PHOTON COMBINATIONS

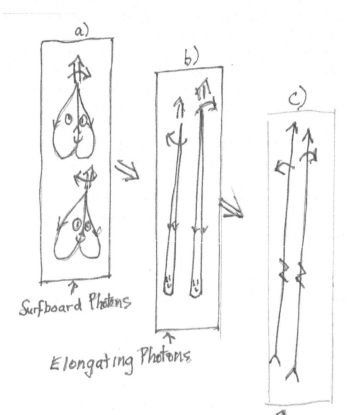

a)

Surfboard Photons

b)

Elongating Photons

c)

Mature Photons
with Polarized Spins.

Different Photon Models

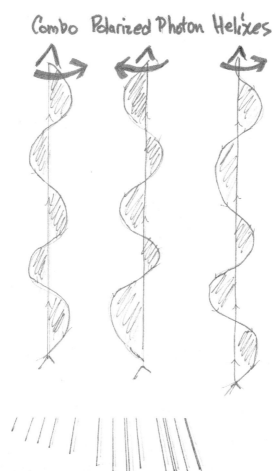

Combo Polarized Photon Helixes

PHOTON COMBOS CONTINUED

Photons spin together, and as more photons combine, the combination grows more forceful, with each photon adding a frequency bump to a long, spring-like helix. The photon helixes above, the favored design, satisfy all the criteria for combination, violin, or single wave concepts—*all of the criteria*. And most important, they are more aesthetically graceful. This is not totally unimportant. Einstein always preferred a graceful universe—as have other great men and women. Apologies if many of the drawing don't give the PS universe the grace it deserves.

If all photons were the same original length, the concept would be even simpler; this should be true, but who really knows? And in some cases, the length might be quite long, hundreds or thousands and thousands of meters. The width, of course, must be close to zero. We will focus more on spinning photons, and why they spin together, in the chapter on magnetism and the electron.

BUT NO MATTER WHAT THEIR ORIGIN, OR WHETHER THEY ARE WAVE OR COMBINATIONS:

At this point, please simply accept that photons exist in a stable form in the PS universe; they all possess wave movement from head to tail, frequency or number, right and left hand polarity, and assist each other to move forward directionally.

And remember, these photons completely fill up the Photon-Storm Universe; they move shoulder-to-shoulder allowing no true vacuum anywhere.

By adopting spins that assist to put them in the most stable state possible, God-photons create a universe much like our own. Backward spin/waves on their exteriors allows photons to move foreword directionally; and clockwise and counter-clockwise spins allow them to easily move together in the same direction and over and under each other without conflict. (See next picture)

WHY THE PS UNIVERSE IS CONTAINED?

More important, since photons in the PS universe require each other to spin forward, the PS universe must have a defined edge. Why? Imagine the universe as a swarm of long fast snakes moving all directions at the speed of light by pushing against each other. But, when a photon from this swarm reaches a true vacuum, it has nothing to push against, and thus can go nowhere; it changes back to directionless energy ball or turns around, spinning back into its home universe.

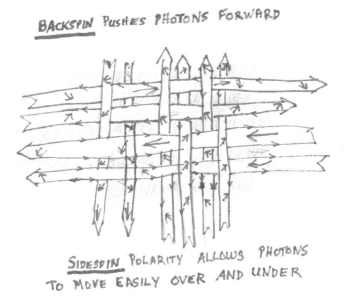

BACKSPIN PUSHES PHOTONS FORWARD

SIDESPIN POLARITY ALLOWS PHOTONS
TO MOVE EASILY OVER AND UNDER

PHOTON POLARITY
The photons above would actually be touching at all edges,
twisting helixes with clockwise and counter-clockwise rotations, long,
backwards spins pushing them forward, smiley-faces, moving all
directions with equal force. [Difficult to draw all that with a crayon.]
If an artist is reading this, please help.

POLARITY OF PHOTONS

Photon polarity is very, very important to increase stability in the PS universe. (See previous pic.) Two photons moving the same direction must adopt clockwise and counter-clockwise rotations to best move together forward. Such polarity also allows photons to weave over and under each other to minimalize sideways interaction. Without such sidespin they would be constantly bumping into each other and would be unstable.

This polarity is shared by electrons and may make a big difference in magnetism, which will be discussed later.

IN CONCLUSION:

A photon, of whatever design, possesses these attributes:

VELOCITY

SPIN – Right or Left, plus or minus ½

FORWARD MOMENTUM – Which equals FREQUENCY or NUMBER of photons, depending on design.

AMPLITUDE -- INTENSITY , which is actually a count of the number of separate photon bundles, or number of separate photons.

IMPORTANTLY: These attributes continue as photons combine into electrons, protons, and universes. So the math is consistent and easy.

CHAPTER 9 -- HOW DOES A PHOTON STORM CREATE MATTER??? THE ELECTRON

Actually, the process of creating matter in the photon-storm universe is quite simple, much simpler than in the standard universe.

Photons simply become incompatible with their PS environment. At a certain configuration, a mega-photon—violin string or combination--must create friction with its neighbors. This semi-particle begins to reflect fellow photons or absorb them when it is struck. The neighbors must treat this mega-photon exactly like a particle trying to muscle through a sea of smaller, lighter, fellow photons. At a certain point, the spin assistance regular photons receive from their neighbors begins to twist a mega photon around. Its frictionless surf through the heavens ends and it is now a different particle, slower than light, reflecting the storm and thus acquiring mass and a credit card.

COMING TO A POINT

In the PS universe, re-configuration must be the key. If photons are directed to the same exact spot—*and if they are large enough they produce magnetic fields to guide them to that spot*—they would unite forever, pushed together by the photon storm. Such point particles would be extremely stable because the force of the storm would be pushing them together. [See pic., p 65] And the S force is, incidentally, the strongest force in the universe; at least it is in the PS universe.

What would cause such a union? Well, if a gamma ray strikes a proton directly, it could be re-configured and much of its energy forced to a specific spot. The high-energy photon could be focused into a particle or two. This process has been observed in the standard universe. Note, the area of a proton is almost zero. Such is our favorite model; and what works for the electron may later work for the proton and neutron.

Mega-photons might also be reconfigured by strong magnetic fields, by interactions with fellow mega-photons, or by running a stop sign and bumping into the edge of the universe. A black hole point is also the perfect place to create point particles.

But other models are also possible and are included if only to create a communal thinking process. The reader should decide. And if the reader has new ideas, please tell us. We are as curious as anyone.

But why would super-colliding-photons ever be formed initially? As we know, photons avoid each other quite easily. Perhaps in the early universe, huge magnetic fields twisted god-photons into huge energy bundles and these degenerated into matter. Or perhaps, in the Photon-Storm universe, natural permutations caused photons to occasionally combine, and over time, an incredible amount of time, enough permutations provided all the matter the universe required.

Or perhaps, an earlier Photon-storm universe was so crowded; the creation of high-energy matter was easier than presently. The edge of the universe scenario is not total fantasy; photon matter may have been focused to points by hitting this early edge.

Or alternately, since the photon storm is a contiguous unity, some elemental force, just like sound waves moving though air, could have condensed parts of the storm in a powerful explosion or collapse. If this loudspeaker condensation was powerful enough, matter could have been formed as the photons pushed together in huge waves of creation.

Now some people might wonder why all this matter could not have come from the explosion of a black hole or super star or big bang. It could have happened that way; and probably did. But black holes and stars already consist of matter or of something that was once matter. We are trying to view a time in the Photon-storm universe <u>long before the creation of any matter</u>. Having a black hole around would be cheating.

A universe of pure energy probably came before anything else, and matter probably came later, at least this seems reasonable in the Photon-storm universe. Little things came first and gradually grew bigger, rather than the other way around.

Energy turned into God-photons, photons combined, requiring—or not--outside energy to do so, grew too energetic and reconfigured until they eventually became matter—at least, this may have occurred in the Photon-storm universe. Black holes came much later. Anyway, that's our story and we're sticking to it, until the reader gives us a better idea. This is a bottom up approach.

SO HOW WAS MATTER CREATED _BEFORE_ THE BIG BANG?

Depending on the unknown characteristics of the Photon-Storm Universe, the initial particles could have been tiny or large enough to break down into smaller particles. If the first matter particle to particulate out of the Photon-Storm was something like the Higgs Boson, a huge particle, then the resulting breakdown would account for many if not all of the known smaller particles today, including quarks, protons, electrons, and so forth. But that easy scenario may not have occurred in the Photon-Storm universe, particularly if the energy level was not the same as in the Big Bang.

In the PS universe, pre-electrons differ from regular photons only in that they have attracted the attention of the photon storm. They can no longer move like their fellow photons; their slower velocities segregate them. The storm may also pummel them without mercy.

The assumption is, such terrible treatment caused these semi-particles to gather together in adolescent gangs, and focus their attentions at specific gathering points. These pre-electrons searched for a point of unity. Some found it; others may still be searching as neutrinos.

The reasonable possibility exists that a number of odd still-nearly-massless particles exist throughout the PS universe, not yet possessing mass but no longer photons, difficult to detect due to their size or elusive nature, but ready to change in a moment's notice.

Assuming again that creation in the Photon-Storm universe started at the bottom making the smallest particles first and working upwards, this is the simplest scenario. If free electrons were created first—apparently the smallest stable particle yet measured--they could have been used as building blocks to create everything else. **Free electron building blocks seem an excellent idea to start.**

Accordian Photon to Matter Particle

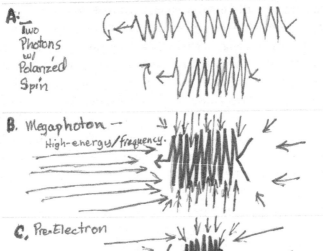

A:
Two
Photons
w/
Polarized
Spin

B. Megaphoton —
High-energy/frequency.

C. Pre-Electron

STORM ↗

VIOLIN MEGAPHOTON TURNING INTO PRE-ELECTRON.

A STRONG FORCE

Using imagination only we can construct a wide variety of pre-atomic particles from poly-photons and pre-electrons. Far too many possibilities exist. We can only consider a few.

To build an edifice, something sticky and strong is necessary to mortar stuff together. In the Photon-Storm universe, this strong stuff can only be conjoined spin and the photon storm.

MATTER: THAT'S THE POINT!

The cleanest electron design, and one that also suggests the nature of matter, is the point design. The point design requires photons come together at a specific point, tying their energy together through the energy of the photon storm. [See following pic]

The following drawing pictures a pre-electron created when several photon bundles or v-strings move to the same spot. This infinitesimally small point becomes matter. A particle with an exact center may call itself an electron. But if the point is wobbly, the matter cannot express itself well and stays in its pre-electron neutrino state.

A neutrino lives between a photon, an accumulation of photons wrapped together with no point, and an electron, a photon bundle that may have achieved point stability. The point-electron has extreme stability because the forces of the PS storm all push it to the same configuration and simultaneously oppose its dissolution. The point electron may consist of as few as three united photons—tiny. Weird, huh!

Such a configuration may attract Bourne-fluff, which would explain its mass. We will discuss the highly scientific fluff later.

The Point Electron

Photons

•Point•

Pre-Electron

Direction

How does a pre-electron turn into an electron? In the following picture the pre-electron runs into a magnetic field and through tough love may be squashed into an electron.

Pre-Electron —
Point Design

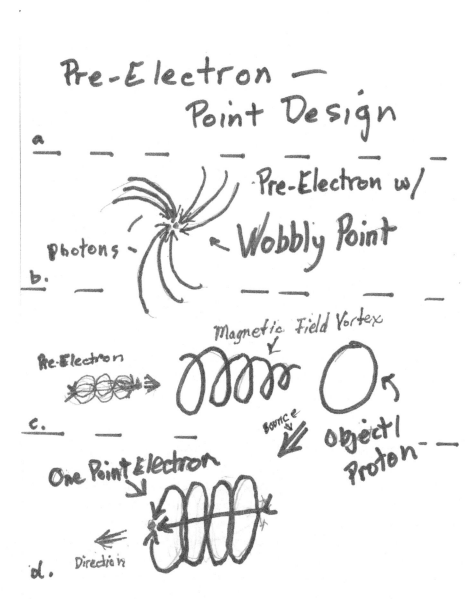

a.

Pre-Electron w/

Photons —

← Wobbly Point

b.

Magnetic Field Vortex

Re-Electron

c.

Bounce

object l
Proton

One Point Electron

d. Direction

In the past pictures, photons chase an infinitesimally tiny point. They have difficulty finding it until a magnetic field (drawing c) bounces them off an object, a possible proton, and when the photons realign they become a pre-electron united around a point called matter.

The point of matter is only material because it—as opposed to photons— bounces the great photon storm and thus acquires a tiny, tiny bit of reflectivity or mass. The new particle may even create mass without occupying any measureable space by interacting with the storm.

PHOTON FLUFF: Once created, the point may assume a life of its own, attracting additional photons or having its configuration and mass changed by collisions or magnetic fields. Such points would possess clockwise or counter-clockwise spins or no spin, and if they combine in sets of three, they would vary depending on whether a or b photons—spinning clockwise or anticlockwise--occupy the center position. Thus 6 configurations are possible in sets of three. A set of four could configure in three ways, positive, negative, and neutral. A set of five could configure in six ways. In general, odd numbers could create six configurations, and even numbers could configure in three designs.

Scientists in the classic universe have a very-rigid concept of mass and consider it to be an inviolable aspect of matter. In the PS universe, mass is a product of configuration. Mass can collect on a configuration in the same way ice crystals attach to a snowflake—we can call this, Bournefuzz.

So the masses of such pre-electrons could vary widely, but spin and charge would be more stable. Intuitively, larger configurations would be possible, but more difficult to construct and less stable.

COMBINATION ELECTRONS

Assuming the skin of photons can join tightly their frequency folds like tiny gears, and assuming the electrons, like photons are quite malleable, the more tightly photons are wound, the more they stick together and the smaller they become. Cubic area means nothing at atomic level, so we need new vectors.

Fe=[(S)/(e)](#e)((Ra)(Rf). Total storm force pushing together an electron particle equals: S force per electron, times total absorbency, times reflectivity. The mass of an electron equals (S), the storm force times the total reflectivity. M=(S)(Rf)

We assume the mass/force of a *compound* electron is related to the number of photons in the electron. Since in the PS universe, electrons are composed strictly of photons, we know there must be a relationship. The formula would be (Ph#), number of photons, times an interacting constant of some sort. We may know roughly the # of micro photons in an electron, since the energy of the electron has been roughly measured. Dividing the energy of an electron by Planck's constant might give us roughly the number of god-photons making up an electron in the compound model, or it may be irrelevant.

Unfortunately, the exact relationship between massless mini-photons and mass cannot be easily measured. The reflective mechanism may depend on variables not yet considered, thus the need for an indeterminacy variable.

Finally, the configurations and therefore the masses of pre-electrons in the PS universe are in constant change. The electron configuration seems to be quite stable in spin and charge, but less so in mass/energy. Electrons constantly absorb and expel energy as photons and in the process may change their energy/mass states.

Best to focus on measureable items, including spins and products of spin, such as charge.

So save this for later.

In theory then, for a compound electron:

An electron in the PS combo-universe is the sum total of the god-photons it contains and the varieties of its spins. It's mass and the forces holding it together are products of the photon storm, photon #, energy level, electron reflectivity and absorbance—all of which may vary.

And electron, exactly as a photon, has these characteristics:

VELOCITY

SPIN – Right or Left, plus or minus, AKA CHARGE.

FORWARD MOMENTUM – Which equals FREQUENCY or NUMBER of photons in an electron, depending on design.

AMPLITUDE -- INTENSITY , which is actually a count of the number of separate electron bundles, or number of separate electrons.

CHAPTER 10 – POSSIBLE ELECTRON MODELS

In the standard universe, photons and electrons are nearly impossible to measure—they appear to be tiny points nearly infinitely small. But in the Photon-Storm universe, which takes place in our imagination, photons and electrons can occupy a great deal of phantom space. They don't occupy all this space at one time, but they occupy it for purposes of visualization.

For example, in the Photon-Storm universe, an energy ball will probably fill up whatever space it is in, and a photon will do the same. An electron will _try_ to fill up all the space available, but it is enclosed by the photon-storm and its own connections to itself, forcing it to behave more modestly and focus on a tiny point.

So, again citing the formula of random probability, we can visualize the electron in several stable forms. The sphincter spinner is one of our favorites. Yes, the probability of this form is unlikely for long; but like its namesake, a sphincter design allows for maximum expansion and diminution. The sphincter design can easily absorb passing photons when stretched by an electric field and emit them later when relaxed. Of course, it would actually be long and spear-like to move forward directionally. Note its backwave moves it forward; it's side-spin creates a magnetic field.

THE (b) SPHINCTER ELECTRON

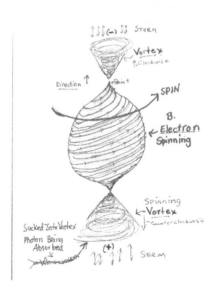

Just like a PS photon, the (b) electron pictured in the past page moves directionally in opposition to its outer spin. The electron also spins at its equator clockwise in relation to its direction of movement. This spin creates a twisting vortex as the photon storm moves through and around it. This twisting of the photon storm translates into a tornado force that attracts and repels. In the case of the (b) electron, the charge in line with its directionality is minus (-).

A photon that moves from a high energy to a lower energy must release the captured vortex, in the form of a photon, in the amount of the energy lost.

An (a) electron (following) spins in a counter-clockwise direction and causes the opposite magnetic field (+) while moving in the same direction as a (b) electron. This drawing has a more reasonable design.

Choice of plus or minus in relation to spin is quite arbitrary. Note that both (a) and (b) electrons in the PS universe have both positive and negative anodes, depending on whether they are approached from the top or bottom. The charge in line with the electron's directionality determines the electron's primary characteristic, whether it moves toward a positive or negative anode.

When clockwise and counter clockwise spinning photons meet, they pull each other together.

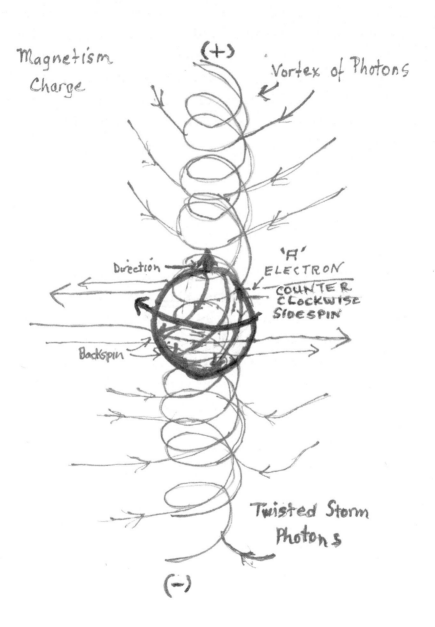

Magnetism
Charge

(+)

Vortex of Photons

Direction

'A'
ELECTRON
COUNTER
CLOCKWISE
SIDESPIN

Backspin

Twisted Storm
Photons

(−)

Bear in mind, the previous (a) electron—a possible anti-electron-- while apparently solid, is mostly empty. The matter that constitutes an electron is never everywhere in its configuration at the same time; in fact, very little of its space is ever occupied at any single moment. Also recall Einstein, "Space is infinite in all directions, without and within."

Thus millions of tiny god-photons can pass through and over an electron unscathed but for being twisted. Those from the middle United States recognize that twisted winds, such as tornadoes, can be extremely powerful. The photon storm is twisted and concentrated in both directions in line with the electron's directional movement in vortexes that attract and repel.

Polarized photons crossing the electron at its equator assist to keep the electron spinning at near the speed of light. In addition, its backwards spin has been twisted due to polarization to not only push the electron forward, but to spin it. (See previous picture)

A (b) electron would spin in the opposite direction and have the opposite directional charge. The (b) electron is the common electron usually found in the standard universe. We have arbitrarily given counter-clockwise spins to (a) electrons and clockwise spins to (b) electrons. The opposite may well be true. We only know they both must spin to create magnetic fields. So, again arbitrarily; (a) counterclockwise spins have been assigned a positive directionality; (b) clockwise spins have arbitrarily been assigned a negatively charged directionality. Both particles have both positive and negative anodes. As particles grow to quarks, their spins and directionality should match those of their electron parents, as electrons match those of photons.

We apologize if these spins are not consistent in the pictures. We get confused easily.

CHARGE AND DIRECTIONALITY

Note in this model, each electron has both a positive and negative anode, but only the charge in line with the electron's directionality gives it its personality. In the following picture, a (b) electron 's direction is aligned with the negative. The (a) electron is aligned oppositely moving toward the positive vortex. If the two electrons were to meet head on, they might well unravel.

THE TWIST ON SPIN

If electrons are forced to spin in polar directions as they are created, why do they continue to spin when they are on their own? Or rather, since electrons a and b are created of the same materials, why don't they simply relax and return to their original states?

The answer may be surprising. Perhaps the original spinning wires of elemental energy retain some rigidity. Perhaps, once they are twisted into a shape, they require energy to be re-twisted into another shape?

And equally important, the forces that shape a particle, the photon storm, continue to shape it. Though an electron may appear to be on its own, it never is. It is *always* shaped and serviced by trillions of invisible god-photons wherever it goes, so its shape will remain stable as long as the field around it does not change.

Note: If spin could be manipulated on electrons, anti-electrons might be produced. These would release a significant amount of energy if they collided with standard electrons in the PS universe.

The following drawings are two electrons in the helix form. The helix form satisfies criteria for either combo or simple wave photons. At the center of the electron are high frequency photons all attached to the same mini-point, and so condensed they have attracted the attention of attached fluff and the photon storm; this attention has turned each one into a particle with a tiny bit of mass. The two electrons are spinning in opposite directions. Spin twists the trillions of storm photons moving over the electrons to create powerful twisting vortexes, which create charge. Arbitrarily the (b) electron spinning clockwise is assigned a negative charge in line with its directionality. The (a) electron spins counter-clockwise and has been assigned a positive charge in line with its directionality..

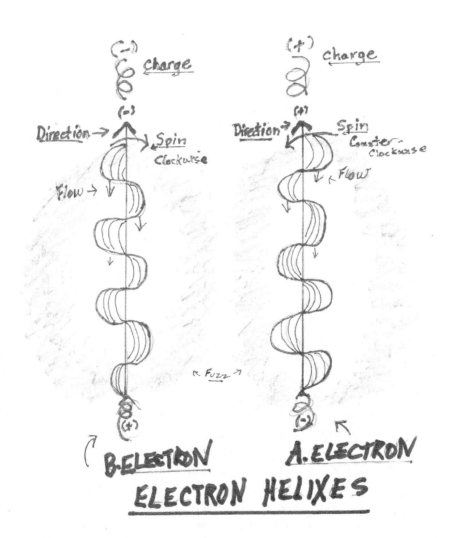

ELECTRON HELIXES

CREATION AND DESTRUCTION

Two combo electrons, A and B, can combine in many ways side to side, united by spin and with their charges partially aligned. However *directly* head to head they might untangle, pulled together by opposing charge and aligned spin, and smashed by their opposing directionalities.

OTHER ELECTRON DESIGNS

Another weird put plausible electron design would be an accordion configuration. This design begins with a single violin string photon, a long, long vibrating pin. As this violin photon—also possessing backwards spin to push it forward—gains energy via increases in frequency, it eventually loses its ability to pass unimpeded through the storm. It begins to reflect photons, rather than having them pass by. And it may even begin to absorb fellow photons, trapping them physically in its accordion folds.

The Vstring and the wadded combo design do not require point stability. The other designs included in the next page benefit from point stability. The originate as single points and then attract a great deal of fluff energy.

The picture is a little messy, particularly if the trapping scenario is true, but the end result a quite similar electron, a very stable configuration, divided into clockwise and counterclockwise spins.

The mass of an accordion electron would depend on its total energy and on how many other photons and semi-particles it attracts, if it does this.

These regular electrons would be divided into left handed and right handed spins giving plus or minus charges in line with directional movement. An electron with no spin would have no charge.

All these electron would have a configuration spin of 1/2 exactly like an electron in the standard universe. (See model)

The violin design also absorbs the simple wave design, the design in which photons are quantum wave functions without pre-history, a plausible design which, if true, would simplify the situation and act similarly to the other configurations.

As in the standard universe, frequency increases and wavelength decreases as particles are created and grow larger. This increase in frequency and mass and decrease in wavelength continues to be true in the quark stage and beyond.

Combinations of violin stings at points would function the same way as combinations of god-photons, but the process would be even simpler.

Possible Electron Configurations

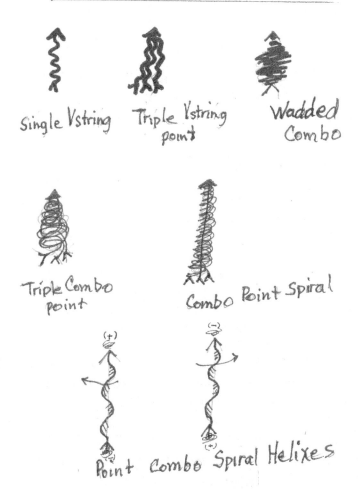

Single Vstring   Triple Vstring point   Wadded Combo

Triple Combo point   Combo Point Spiral

Point Combo Spiral Helixes

Both configurations—the Vstring and the combination--can absorb and emit other photons; and both configurations react with the passing storm via spin to create twisting storm disturbances of electromagnetism.

MANY OTHER CONFIGURATIONS ARE ALSO POSSIBLE.

These models just seem somewhat reasonable and seem to work for now.

POLARITY, VERY IMPORTANT

The following *possible but not probable* pre-electrons may have been created in two varieties, with a clockwise and counter-clockwise outer spin compared to an inner spin which maintains the integrity of its photon and energy ball components, call these A and B sidespin. These polarities correspond to their stable states as mega-photons.

However, besides having equatorial spin in clockwise and counterclockwise directions, these pre-electrons may have polar wobble; this polar wobble is what makes them susceptible to being acted upon by the Photon-Storm by reducing their forward movement. The total spin possibilities of stable pre-electrons then increase—'A' pre-electrons counterclockwise, 'B' pre-electrons clockwise, pre-electrons with a polar wobble north to south, and finally, pre-electrons with neutral equatorial spin. The total is at least three pre-electron varieties distinguished only by their spins and up to five if a polar wobble is a factor.

Also in consideration, such pre-electrons may exist in unfinished conditions, with divided spin or mass, half (a), half (b) combos, or with photon attachments.

The (A) and (B) varieties, however, would appear to be by far the most stable.

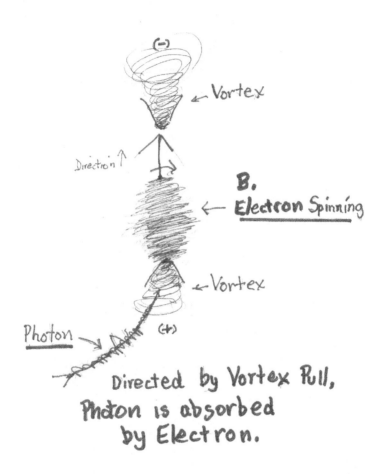

(-)

← Vortex

Direction ↑

B.
← **Electron Spinning**

← Vortex

Photon → (+)

Directed by Vortex Pull,
Photon is absorbed
by Electron.

Above,,a (B) electron absorbs a photon pulled in by vortex. Sidespin in a particle twists the storm to create charge at the poles. Since (a) and (b) electrons have different sidespins, their charges are reversed.

The idea is, spin in an electron particle must be similar to that imagined in polarized photons, but electron-particles lose some of their directionality and are large enough to attract, bounce, absorb, and emit other photons—and they are also large enough to twist up the storm creating magnetic vortexes.

Frequency increases; wavelength decreases, as particles grow larger.

CHAPTER 11 -- MAGNETIC CHARGE DUE TO SPIN

A REVIEW: ROTATION FORCES ELECTRONS TO EXERT A MAGNETIC FORCE. THIS FORCE IS IMPLIMENTED VIA THE PHOTON STORM. TRILLIONS OF COLLLISIONS OCCUR EACH SECOND WITH MINI-PHOTONS TOSSED BY A SPINNING ELECTRON IN OPPOSITE SPINNING VORTEXES DEPENDING ON THE ELECTRON POLARITY. EQUATORIAL SIDE SPIN CAUSES TORNADOE-LIKE AFFECTS AT AN ELECTRON'S POLES.

ELECTRON VORTEXES CAN NULLIFY EACH OTHER WHEN THEY ARE AT ODD ANGLES. BUT IF THEY ARE ALIGNED WITH THEIR POLES IN A SIMILAR DIRECTION, AS IN A MAGNET, THEY WILL COMPLIMENT EACH OTHER AND INCREASE THE AFFECT.

IF ENERGY STRETCHING AN ELECTRON FIELD SUDDENLY DISAPPEARS, A VORTEX WILL BE RELEASED AS A PHOTON IN THE AMOUNT OF ENERGY EQUAL TO THE LOST STRETCH.

IF LIKE POLES ARE ALIGNED THE SPINS WILL OPPOSE EACH OTHER AND THE RESULT IS REPULSION, AS VIEWED IN THE PICTURES BELOW. OPPOSITE POLES CREATE VORTEXES THAT ATTRACT.

AN ELECTRON WITH NO EQUATORIAL SPIN WOULD NOT HAVE ANY POLAR MAGNETISM BY THIS MODEL. A PARTICLE WITH NO SPIN OR ALL ITS SPINS BALANCED WOULD HAVE NO CHARGE.

WE WILL LATER SEE HOW VECTOR COMBINATIONS GIVE CHARGE VALUES TO THE PROTON AND NEUTRON (Page 98).

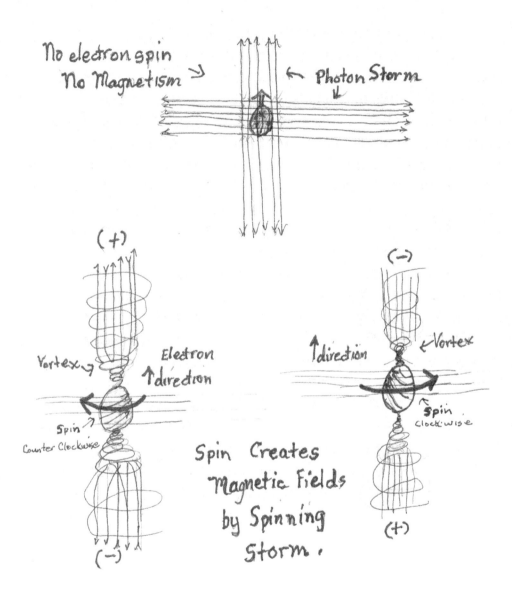

No electron spin
No Magnetism →

← Photon Storm

(+)

Vortex →

Electron
↑ direction

Spin ↗
Counter Clockwise

(−)

↑ direction

← Vortex

Spin
Clockwise

(−)

(+)

Spin Creates
Magnetic Fields
by Spinning
Storm.

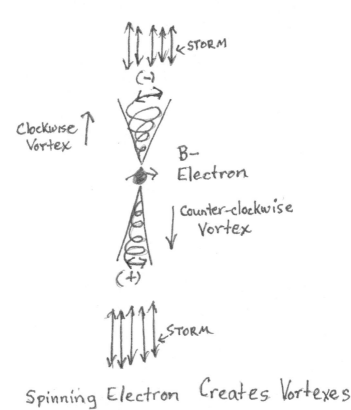

Spinning Electron Creates Vortexes

So to repeat, because of spin at their equator, electrons create a magnetic field by twisting together the photon storm moving over them. Opposing spins attract; matching spins repel. Electrons also have backwards wave spins like a photon. Backwards spin on their exteriors helps electrons move forward through the storm, like photons.

Photon Vortexes Twisting
in opposition —
Attract!

Note: Photons pull each other
in the opposite direction to their
own movement.

Electrons

(−)   (+)

Attraction
of Opposing Photon Spins

SIDE BY SIDE FORMATION: Pre-electrons can be divided into 'a' and 'b' by
their clockwise or counterclockwise rotations. 'a' electrons can match side-by-side

spin and direction with 'b' electrons; however 'a' electrons can not match side-by-side with other 'a' electrons; and neither can 'b' fit with other 'b' electrons—*unless they are moving in opposite directions.* See the models on page 61. Their spins would clash.

So, *if the side-by-side model is correct,*--A BIG IF WHICH IS NOT OUR FAVORITE--the electrons would combine a-b-a-b in strings or matts moving the same directionally. (See pictures pgs. 95, 96)

ANTI-MATTER

The concept of anti matter takes us all the way back to energy balls. If these theoretical balls can spin inside out and outside in, and if the two spins cancel out, then two universes, one of matter and the other of anti-matter might be possible. Perhaps stability has separated these two universes or perhaps they have cancelled out leaving only a tiny bit of matter to stock our universe. Or perhaps, this scenario never occurred.

HOWEVER, in the PS universe, particles can come together in ways that cause both of their spins to untangle, creating the same effect as if a particle had met up with its anti-particle.

FOR EXAMPLE: If two electrons of opposite polarities come together side to side they could combine—however, pushed tail to tail or directly head to head they might unravel releasing their energy.

A magnetic field capable of sorting electrons by their spins and recombining them in the PS universe might produce a great deal of energy.

WHERE DID ALL THE ANTI-MATTER GO?

Some physicists still wonder why our standard universe is not equally made up of matter and anti-matter. Perhaps only 'a' or 'b' electrons dominate, since one or the other has been returned to energy by mutual annihilations. In the PS universe, the answer to where the anti-electrons have gone is *possibly* that they are tightly wrapped up in indivisible hadrons or protons. So, anti-matter may never have disappeared in the PS universe, it may be simply tied up in the process of matter. (See pics. 98)

Actually, such a configuration seems logical according to the stability model.

(Or alternately, an anti-matter universe might exist, or opposing energies exist inside out in photons, or...)

REVIEW

Reviewing everything so far, the concepts extend from the original premise of a photon storm. Each extension offers an opportunity for error.

Ideas began with simple elemental particles that combined in very simple ways to create a universe of photon energy.

The spin/wave concept demonstrated how energy moves in the PS universe and why the energy is contained.

This photon energy storm then eventually formed high-energy photons. Magnetism assisted photons to spin together or meet at the same exact spot, and these spots grew so unwieldy by reflecting and absorbing fellow photons they had to turn into matter. Point electron construction created stability.

The reflectivity of matter produced *gravity*. And the absorbance and spin of matter defined the *weak forces.* The spin of electrons and protons twists the photon storm to create magnetism. The ability of electrons to absorb and emit other photons allows us electronics and our smart phones in the PS universe. The photon storm provides the *strong force* to push photon bundles into electron and proton points of matter.

At each step concepts build on previous conclusions, and so the risk is high the ideas could make a wrong turn somewhere and the error might magnify. If wrong turns happen, we expect the reader to inform us so we can improve the situation together—or toss out the whole concept and take a nap. Naps are nice.

The advantage of a simple model is that everyone can play with it.

MATH

And incidentally, math in the PS universe is simple. The photon design with spin, directionality, #a or frequency, #b or magnitude, times Plank's constant will continue through the electron and photon designs without much change—except for the addition of fluff, which requires more explanation.

CHAP 12 – LARGER PARTICLES, AND HOW THEY GET TOGETHER

TWO WAYS TO PROCEED

(Or skip to the favorite design, pg. 98!)

THE VIOLIN STRING

We have two models for the photon in the PS universe—the violin string model and the combination photon model. Each model generates a slightly different concept for creating larger, stable particles.

The violin string model is quite simple. Different frequency modulations on a string can create different particles at different energy levels. The electron, for example, exists at one energy level of a violin string. Neutrinos, quarks, and bosons are simply other frequency modulations.

Combinations of strings modulating cannot be excluded--but in theory a single violin string may extend to extraordinary mass and energy. The creation question is thus answered: All matter is composed of violin strings at extraordinary tension vibrating through multiple dimensions to create the world, as we know it.

In the standard universe, such a definition is generally accepted as of 2017. Electrons, quarks, protons, neutrons—with all their affectations—are the creations of violin strings (possibly in combination) with wildly frenzied modulations. Larger particles, yet undiscovered, even beyond the Higgs, probably exist beyond our present ability to identity them.

Such a situation may be true in the PS universe as well. Violin string theory cannot be excluded in the Photon Storm domain.

However, since the PS storm concept has elimination the need for extraordinary leaps of mathematical faith in many areas (revisit the storm solution to the double slit experiment of pg. 7), we might look for *simpler versions of particle creation in the PS universe also.*

Most importantly, the photon storm itself should be a major force in the creation effort. A storm of trillions of energy photons on all matter should not be ignored.

In the PS universe, all particles, whether vibrating strings or photon combinations, are pushed together by the extraordinary force of the photon storm. And since, the PS universe is simple we can even visualize the process of putting together building blocks to create large particles.

THE VIOLIN STRING COMBO

The violin string design is easy to envision. How would several violin string particles combine? The model creates a particle that possesses most of the attributes of a photon, plus magnetism and fuzz.

1)   DIRECTIONALITY, the push in one direction by back spinning oars like a photon.

2)   SIDE SPIN, equatorial rotation, clockwise or counter clockwise, possibly a result of twisted back-spin.

3)   MAGNETISM, tornadoes of spinning god-photons that create pulls and pushes of energy at the particle's poles—arbitrarily designated + and -.

4)   FUZZ, *stuff* in the form of god-photons and semi-particles that attach to particles.

5)   # or frequency

To create combinations out of such particles, the designers must keep in mind matching spins, magnetism, and directionality.

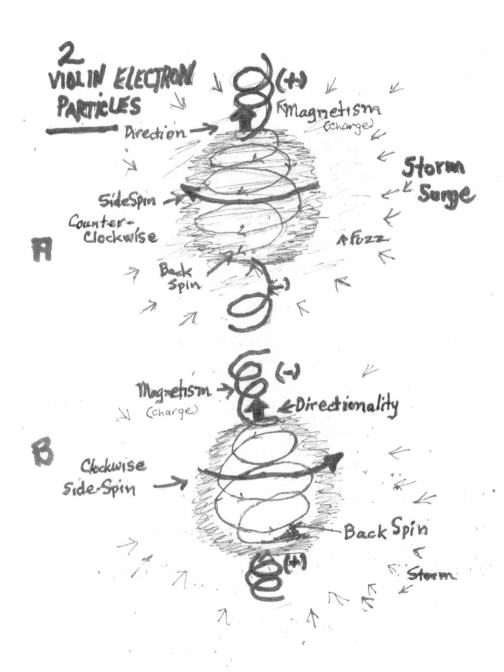

2
VIOLIN ELECTRON PARTICLES

Direction →

(+)

↑ Magnetism
(charge)

Storm
Surge

SideSpin →

Counter-
Clockwise

↑ Fuzz

A

Back
Spin

(-)

Magnetism →
(charge)

(-)

← Directionality

B

Clockwise
Side-Spin →

Back Spin

(+)

Storm

In the previous picture, see two particles created from the violin/accordion model of the photon. The only difference between the particles is sidespin—one particle spins counter-clockwise, the darker one, the other particle spins clockwise. They both generate magnetic fields by spinning the storm surging around them, but with opposing spins their magnetism is also opposing.

MOST IMPORTANT, observe their directionalities. They move in the direction opposite to their back spins, being pushed at nearly the speed of light by their interaction with the storm, similar to how photons are pushed by the storm—but, having mass, they are slowed a bit.

How could such energy particles combine? Suppose they increase their energy 5 or 10 fold respectively and become 2 up quarks and a down quark.
(See pic.)

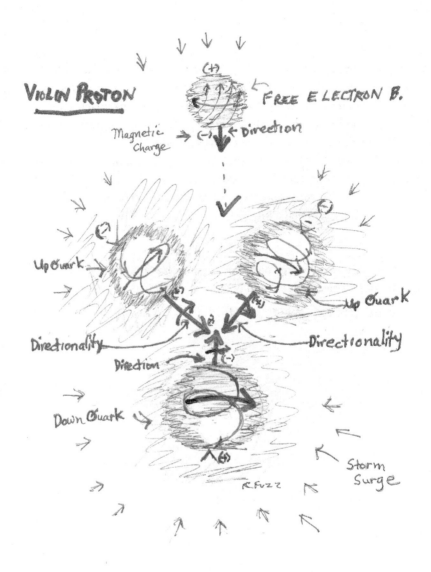

VIOLIN PROTON

FREE ELECTRON B.

Magnetic Charge → (−) ← Direction

Up Quark

Up Quark

Directionality

Direction

Directionality

Down Quark

Storm Surge

R. Fuzz

V-STRING PROTON

See in the previous picture how such particles might combine. The two up quarks would repel each other by spin and charge. They would both have a positive directionality. However, if a down quark spinning contrarily with a negative directionality came from the opposite direction and they met in the exact middle, the spins of the up quarks would be complementary to the down quark. They could join amiably, similar to an oil drill bit.

IF the three particles joined, they would be pushed together by their spins, by the gravity of the storm, by their joined magnetism, and by the total force of all their directionalities. In other words, every force in the Photon Storm universe would be pushing them together!!!

In other words, such a particle would be very, very, very, very—is that enough?—quite difficult to disassemble! It would be extraordinarily stable. Therefore, even if the probability of such particles meeting together were small, if they did meet, they would never part. Eventually such a stable particle might dominate the entire PS universe.

Also note the cozy positive, (+) spot at the bottom of the down quark. Such a cozy corner might easily draw a free electron following its directionality and pulled by its (-) negative charge to attach itself with some permanence. Also, if the entire combination had a clockwise spin via its directionality, its positive path would lead to an electron and vice versa. The electron's attachment would be limited by the spins of the up quarks, (See drawing) which would oppose it, giving the electron the freedom to cavort at times as electrons enjoy doing.

A neutron, composed of two down quarks and an up quark, would have no specific sweet spot for a free (b) electron, but it should attract an (a) electron, were one available, or an a-b-a-b string in the combo model.

FUZZ

Note also the entire proton would inevitably be surrounded by a great deal of fuzz—call it Bourne fluff--associated energy attached or absorbed by the central particle. Such fuzz could add to the Reflectivity (Rf) of the particle and greatly increase its perceived mass.

Note in addition that the single proton might possess a slight directionality in the positive direction—but by adding a free electron, its combined directionalities would balance. However, as a complete particle it could still spin around an axis just to complicate matters.

COMBINATION PROTONS built from ELECTRONS

SIDE BY SIDE: Combination particles, created from combinations of semi-electrons, are *possible* in the PS model, and would go far to explain the perceived masses of assembled quarks. However, they are quite ungainly.

In the pictures on pg. 90, are drawings of (a) and (b) electrons. The electrons differ only in their sidespins. Note each electron also has a direction it is being pushed by the storm, and positive and negative charges caused by sidespin twisting the storm. To use these semi-electrons for building blocks, they must match in conjoined spin, magnetic charge, and directionality.

These side-by-side configurations may be unlikely, but are possible, so they have been included in the examples on the following pages. (See pics)

If we take three counterclockwise (b) electrons and combine them with two clockwise (a) electrons, we have a combination of five semi-electrons with directionality. Call these particles up quarks (see pic.). Take these two up quarks of five electron blocks each and run them into a particle with four (b) electrons and six (a) electrons going the opposite direction, possibly a down quark (see pic. Following, pg), and the result is a balanced mass particle somewhat similar to that of the violin design.

Like the violin model, this combination side-by-side proton model could be very, very stable, pushed together into a tiny, tiny dot by spin, directionality, magnetic charge, and the storm force combined. A great deal of Bourne-fuzz could also attach if this fuzz needed a place to hang out; such associated fuzz could explain the small masses of the quarks and the larger mass of the complete proton as measured in the standard universe.

The tiny proton point would also be much more dense than its original electron building blocks with a modified gravity pull and an increased ability to absorb photons pummeling it.

A danger in the combination spin model is that electrons are in close proximity to electrons of opposing spin, and a small change in alignment might cause the two particles to unravel. In other words, the combination model places electron bricks in close proximity to their possible anti-particles.

Finally, note in the upper right corner a free (b) electron. This electron would be drawn to the combination proton directionally by storm gravity and by its electric charge. It would find several attachment points at positive points between the up quarks or at the end of the down quark. It could match spins fleetingly with (a) electrons, but the spins of other (b) electrons would repel it. Thus its attachment to the proton would be variable and the electron would be relatively free to wander.

(See pic.)

The people in the PS universe should be very thankful for wandering electrons, since they provide them with the benefits of all electronics.

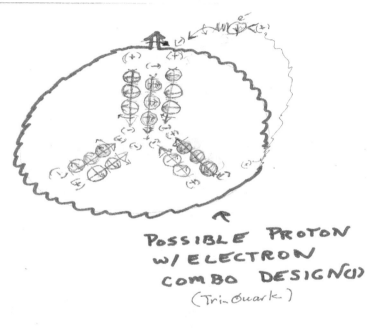

POSSIBLE PROTON
W/ ELECTRON
COMBO DESIGN(1)
(Tri-Quark)

The previous drawing is a combo proton with a side-by-side tri-quark design—an unlikely possibility.

The following drawing is a combo proton/neutron with a bi helix configuration.

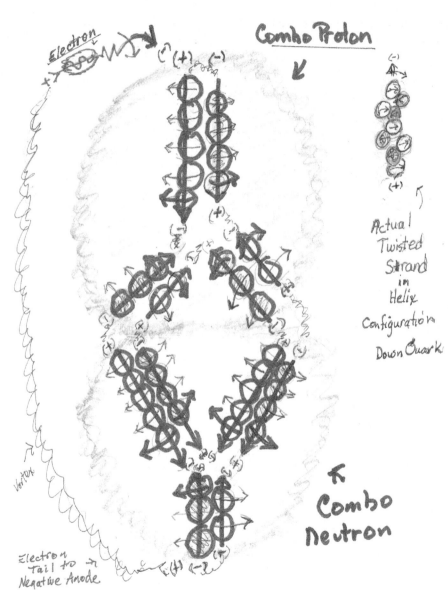

The side-by-side drawing on the previous page has better possibilities. It is built from (a) and (b) electrons: balanced in mass and direction. Two possible up

quarks of five electron mass each, oppose a down quark of nine electron mass. However, as a two-dimensional drawing, it fails to portray the roundness of the final proton configuration. By intuition, the quarks strings should be twisted into helixes. Also in the PS universe, the proton should be round, and so some of the positions of the electron blocks might need to shift to achieve roundness. If these block shifts continue to occur after the proton is stable, then proton decomposition might be possible naturally, though highly unlikely.

Note the storm current within the combined hydrogen isotope is connected through both the proton and neutron with anodes connected. The electron increases the vortex flow by connecting externally. If the neutron were not present, the electron would try to connect to the bottom of the proton. In both cases it is relatively free to seek attachment with a separate atom.

The previous drawing is a simple combination of electron blocks by spin. If it is accurate the model partially explains the mass of quark combinations. Such a combination would be extremely stable in the same way as the violin proton, since it is pushed together by every force in the PS universe. Such a combination would also attract a great deal of Bourne-fluff –as it has been called--from the storm; this fluff, pushing from all sides, would assist to turn the composite proton into a sphere; and it would also add to perceived mass.

Note also: The b-a-b-a-b electron strands would probably assume a helix configuration, because this design is more graceful. HOWEVER...

THE ABSOLUTELY-POSITIVELY SUPER-SIMPLEST PROTON/NEUTRON POINT DESIGN FOLLOWS!!!

The simplest design may be the best. And strangely, this design seems to explain almost everything. In this model up quarks consist of three electrons, two a's and a b; the down quark is only 5 electron blocks, with 3 b's and 2 a's. All the electrons in each quark are directed by the storm to the same exact point, and when combined the configuration would be extremely stable. However, we know the neutron by itself is not stable., with a half life of only ten minutes. Why? In this model, built from three spinning vortexes, perhaps the two down quarks spinning at near the speed of light occasionally clash. If they did, two electrons might be knocked from the spinning down quark turning it instantly into an up quark. One electron, the b electron would be set free, the other electron, an a or anti-electron might be more damaged. It would limp off as an anti-neutrino. *This phenomenon is exactly what is observed in the classic universe, so the model might have some validity.*

Recall, in the PS model spin is the weak force that causes atomic
disintegration.

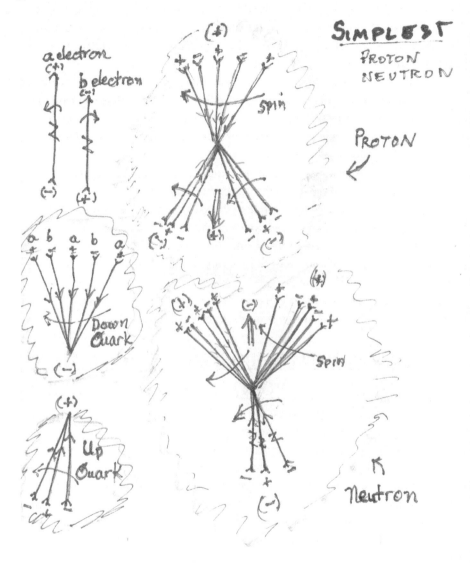

PROTON NEUTRON STABILITY

Note in the previous picture, the positive arms of the neutron would attach to the negative arms of the proton. If such a configuration occurred it would explain the extreme stability of a neutron once is it connected to a proton—the down quark arms are held apart, and thus they can do less damage to each other. The stability is even more structurally obvious in an alpha particle or helium atom (pg. 122).

The previous model might also explain the transition of a proton into a neutron. The proton would attract a b electron, but that electron would not be able to pass by the negative wings; rather it would hang on like a mosquito. However, under extreme conditions, the electron could unite with a positron, creating a neutral point that could make its way past the negative wings to the center point. The proton would then be a neutron. Such a situation would have great difficulty occurring if the proton were already attached to a neutron, see pic. (pg. 121).

QUARK CHARGES EXPLAINED

Assuming the previous simple quark models have validity, then the design explains the various quark charges and their differences. An electron has a simple charge of 1 or -1, depending on its spin. When electrons unite around a point, their charges move at angles compared to the directionality of the combined particle, so the combined charge is spread out and a vector quantity of the original charges. A combination of three electrons would have a smaller vector angle than a five-particle combination and thus a more focused charge—but still not as focused as a single electron.

Once quarks unite as protons and neutrons, the angle of charge may have reached its peak. So combinations of quarks may unite their charges linearly, explaining the positive charge for a proton and the neutral charge for a neutron. To check, these angles may be somewhat easily computed.

(See pic.)

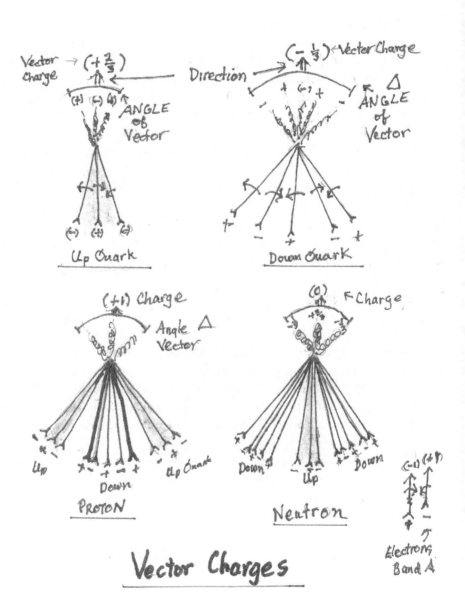

Vector → $(+\frac{2}{3})$ ← Vector Charge
charge

Direction →

$(-\frac{1}{3})$ ← Vector Charge

ANGLE
of
Vector

ANGLE
of
Vector

Up Quark

Down Quark

$(+1)$ Charge

Angle △ Vector

(0) ← Charge

Up

Down

Up Quark

PROTON

Down

Up

Down

Neutron

Electrons
Band A

Vector Charges

THE PAST DESIGN IS THE BEST AND NEAREST TO NATURE!!!

Some will argue: How can a proton configuration made up of only eleven electrons have the measured mass of a proton equal to 1837 electrons? The answer again is FLUFF, BOURNE-FLUFF. The photon storm contributes most of the mass of a proton configuration, so in the PS universe, mass is not a problem.

The possibility of many unstable designs is quite high, some having high masses. The quark design could continue indefinitely. However, once particles join strongly with opposing directionalities as in a proton or neutron, the chance of change is small.

PROTON INSTABILITY

In normal situations, the proton design—combo or violin string—is extremely stable. But at high speeds approaching the speed of light, the forces holding a proton together begin to weaken. (See model) The directionality force would approach zero. Theoretically, if an accelerator could power a proton to the absolute speed of light, that proton might well disintegrate on its own in the PS model.

PHANTOM PARTICLES, The Wo, W-, and Zo and the Graviton

The PS model has some difficulty explaining perceived particles in the standard universe, the Wo, W-, and Zo, for example. These particles are complications in a model that depends for almost all its phenomena on a simple photon.

Fortunately, the large masses of these particles match—with a great deal of creative adding and subtracting—the missing masses in complete photons and neutrons. In the PS universe, such mass differentials are attributed to fluff, a highly scientific designation meaning the amount of photon bundles that attach to stable quarks.

The picture is of a spinning three-pointed jack with massive photon banners entangled in its many spins and charges. These banners are supplied instantly to the particle by the photon storm. By reflecting fellow photons, these happy banners add greatly to the perceived mass of elemental particles. They also add to the necessary roundness of elemental particles.

When a proton is rudely torn apart, these bundles may appear momentarily as *real* individual particles of the photon variety. Without quarks to hold onto, they are totally unstable—able to disappear and reappear instantly from the additions of trillions of stray fellows from the photon storm. Unlike in the standard universe, these unstable particles are absolutely real for the microseconds they exist. No phantoms in the PS universe.

A similar graviton bundle may also be pulled from the storm when the gravity shadow is stretched to the maximum.

HEAVIER QUARKS AND LEPTONS AND MASS

The absolutely positively simplest explanation for heavier quarks and leptons is the most natural configuration based on the electron, or quark design—that being an electron particle united head to tail. In the following picture, an electron has a charge of (-1). When united head to tail to another electron, or a pre-electron, the particles may become a muon. The charge stays the same, since the diameter of twisted god-photons remains basically the same. However, an electron neutrino muon might attract much more fluff than a single electron; this cape of fluff may increase the mass on a muon many times that of a single electron.

Up and down quarks may also combine sequentially, while maintaining the same charge, to create similar particles with greater mass but far less stability.

See the following drawings of electron and possible heavier muon and tau, plus up quark and heavier charm and top quarks.

The drawings are not accurate in that somehow the electron, muon, tau family is struggling to reach the same point; they would not be separate particles combined in a chain, but a single semi-stable particle with very unstable following bushy tails. The same is true of the quark families. They would grow less stable as they increase in size.

PARTICLE FAMILIES

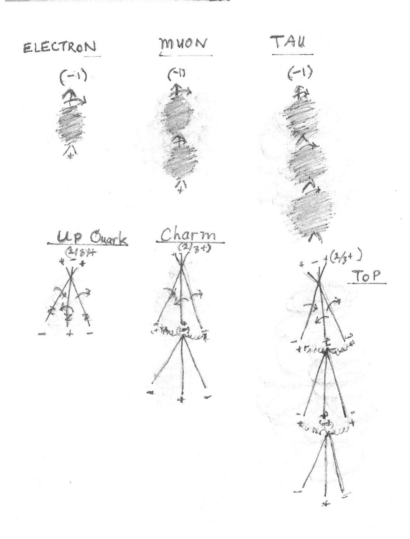

ELECTRON

(−1)

MUON

(−1)

TAU

(−1)

Up Quark
(2/3+)

Charm
(2/3+)

(2/3+)

TOP

Again, in the PS universe, mass is dependent on configuration. Mass accumulates on configurations like ice-crystals gather on snowflakes. This accumulation has derisively been called, Bourne-fuzz. If true, the energy for growth comes not from the particle but from the surrounding photon-storm.

When electrons gather around a point, the point slows down, and each quark attracts a reasonable new crowd. If an up quark gains approximately 203 times its reflectivity in fluff, and a down quark fattens by 123 times, then the weights of protons and neutrons would make sense.

In sequential particles, like the muon and tau, each new configuration would bring with it new points of attachment. If X attachment points attract Y photons in the first configuration, then the second configuration would have XY attachment points, and if Y photons attack to these points, the mass of the second configuration would be relative to XYY. The third configuration might have a mass *somewhat relative* to XYYY, and so forth.

The above concept assumes some continuity in particle construction and mass accumulation; *this continuity may not exist.* In addition, increasingly larger configurations—like the up and down quark families--would be affected by more complicated spin variables. And finally, the general rule is, the greater the mass, the less the stability.

THE ELECTRON STAR??
WHAT FORCE COMBINED SIMPLE PRE-ELECTRONS INTO LARGER PARTICLES?

In a universe with the electron being the largest particle, electrons would eventually group together due to gravity, magnetism, and God's hammer to form electron stars. Electron stars could be the caldron to create larger particles, electrons, quarks, and protons; in the same way hydrogen stars push protons together to create larger atoms.

An electron star would be composed of long beads of a and b electrons. Such a creative star would allow great opportunity for electrons to combine, particularly in their simplest forms.

A black hole or big bang would also be an excellent way to combine matter by aligning points, since in the black hole state, the proximity of all god-photon points would be very close together.

CHAPTER 13 – <u>MAGNETISM, ELECTRON CHARGE, ELECTROMAGNETISM, AND RADIOACTIVITY IN THE PHOTON-STORM UNIVERSE.</u>

By the way, the electron does not orbit the proton in the PS universe; it is actually attached magnetically to the proton with everything spinning. Weird, huh. And fluff energy circulates and creates most of the mass.

GRAVITY PLAYS A PART IN PARTICLE ACTIVITIES

In the following picture of an electron being pulled to a proton by its charge, note the electron is less dense when compared to the proton—at least its energy is freer. The condensed proton soaks up a great deal of pressure from the photon cloud due to its possibly denser configuration. So gravity also plays a part in drawing the electron to the proton.

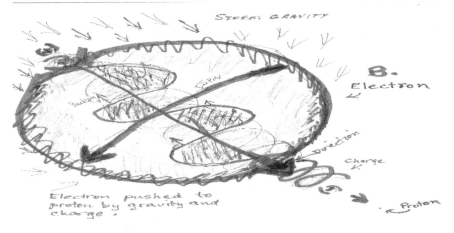

Solo Protons Repel Each Other

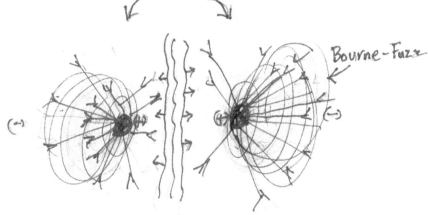

Bourne-Fuzz

(-)   (-)

PROTONS Repelled by Magnetism and Fluff Mass

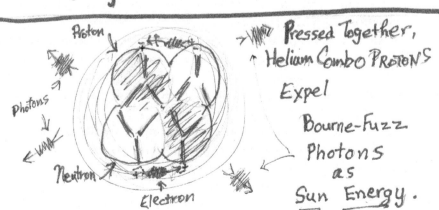

Proton

Photons

Neutron

Electron

Pressed Together,
Helium Combo PROTONS
Expel
Bourne-Fuzz
Photons
as
Sun Energy.

PROTON REPULSION

In the past picture, two protons repel each other. The repulsion is a combination of opposing magnetic vortexes plus opposition by spinning fluff mass, which is most of the proton's mass. In the bottom picture, when two protons and a neutron are pushed together forcefully, they fit neatly, but must expel a portion of their fluff, also derisively called Bourne-fuzz. This fuzz exits as sun energy. The combination also attracts electrons. Neutrons and electrons assist in allowing protons to stay together, forming quite stable combinations (See pic 122). *The entire atom is pulled together by its associated vortexes and the power of the storm.*

In the Photon-Storm universe, the creation of larger nucleuses by the combination of hadrons—protons and neutrons--is reasonably uneventful, a matter of utilizing storm gravity power to force hadrons together once repelling forces have been overcome.

The single force that impels matter to form stars and create new matter—in the Photon-Storm universe--is the photon-storm. The same storm creates stars and the nucleuses of large atoms, keeps the nucleus together, assists in chemical reactions and the structure of all matter, and pushes electrons in their routes.

All these responsibilities may seem a lot, but the god-photon storm accomplishes it all without complaint. It is, after all, a god photon.

IT'S ALL ABOUT THE ELECTRON
Electricity, Magnetism, Electromagnetism

Choosing electrons as blocks to build larger particles may be debatable, but electrons are certainly *absolute stars* when it comes to almost everything else— charge, magnetism, electricity, and electromagnetism.

But, unlike in our standard universe, in the PS universe electrons are seldom left to their own devices; they are always being pushed in some beneficial direction by the parental herding of the photon storm.

Electrons are assisted to find stable unions with atomic nucleuses by the storm.

Electrons actively combine spin with their proton nucleuses and in larger molecules they can attach to more than one nucleus; this electron attachment and storm activity actually helps hold molecules together.

When atomic nucleuses only weakly hold outer electrons, such as in conducting metals, the photon storm can push electrons toward a more positive situation, assisted by the charge created by the electron's continued spin.

When electrons are stretched by a magnetic field, their chinks are happily filled by photons from the storm, drawn in by the electron's powerful vortexes, and when those same high-energy electrons relax, they emit electromagnetic energy in the form of combination photons--all due to the assistance of the photon storm. (See pic on following page)

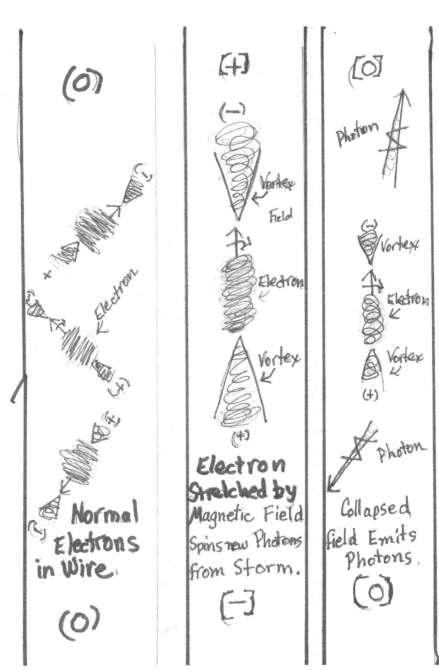

CHARGES, POSITIVE AND NEGATIVE

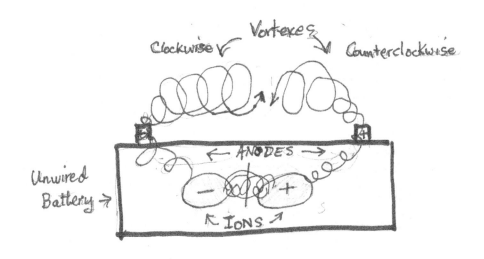

Clockwise Vortexes Counterclockwise

Unwired Battery →

← ANODES →

− +

ↆ IONS ↗

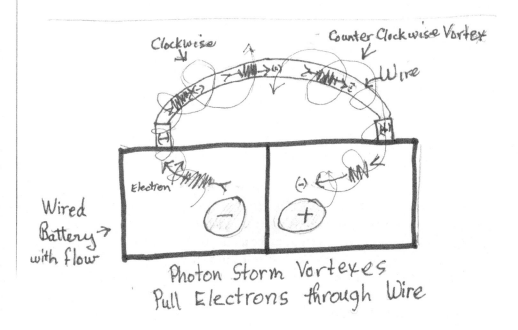

Clockwise

Counter Clockwise Vortex

Wire

Wired Battery → with flow

Electron

(−)

−

+

Photon Storm Vortexes
Pull Electrons through Wire

The photon storm fills the gaps when electrons stretch to higher energy, and when electrons relax to lower energy states, electromagnetic photons are pushed out, giving us light, radio, television, and smartphones.

In a battery, (past page) opposing vortexes stretch between positive and negative anodes. A metal wire allows electrons to flow in the direction of charge to complete chemical reaction. Vortexes of twisted photon storm provide the power behind the electric flow.

MAGNETISM is easily visualized in the PS model (next page). Each electron creates a tornado at both poles of clockwise or counterclockwise photon-storm spins. When oriented in the same direction, these streams combine dramatically moving both directions from pole to pole. Long spinning vortexes of photons create fields we can call positive and negative depending on their clockwise or counter-clockwise orientations. Opposite rotations attract: similar rotations repel. Eventually the positive and negative spins combine.

The flow of god-photon storm vortexes is blind to changing materials, moving easily through air, unlike the larger electrons.

If a wire coil moves around the magnet, electrons in the coil, controlled by their charge, are pulled through the wire by the vortexes they encounter, so that a current flows in the coil. The cause of both the current and the perceived magnetic lines, is the power of the photon storm when twisted into vortexes.

Recall, in the PS universe, god-photons pull against each other (pg. 37).

In stars and planets the situation is similar to a single electron or a magnet. Stars and planets are spinning rapidly; they also possess liquid, flowing parts possessing polarize ions. Spin of a star creates an actual electronic flow of storm-pushed high-energy electrons and twists the storms of God-photons into visible spiraling tornadoes. The collapse of these photon tornadoes releases tremendous energy from associated ions and electrons.

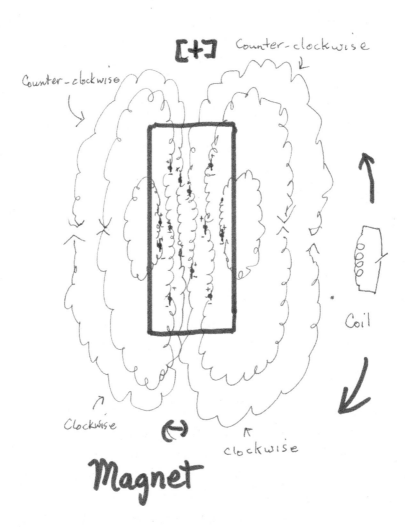

CHAPTER 14 -- BUILDING ATOMS FROM QUARKS IN RATIONAL MODELS

To repeat: As the PS model continues to build on itself, it grows ever more variable. In other words, any one of the previous designs may be inaccurate, and this inaccuracy will grow as the model extends to other areas, such as the creation of atoms or molecules.

So it is important to repeat that no claim can be made of accuracy, and that the reader must be free to reframe the model whenever something seems out of kilter. Proceeding further is probably increasingly reckless, but to stake out some initial possibilities seems reasonable, particularly since the architecture appears to explain some phenomena. The point is, a misstep at this point should not deconstruct the entire previous argument. If the reader encounters something irrational, move forward with your own concepts. The model welcomes any input.

Again, the premise arises from the stability concept. If we start with energy balls, imagine photons growing from these balls with directionality, and further imagine how these photons can grow in energy and combine into particles, the sole criteria is stability. What is stable will continue to exist; what is not stable will disappear. And we project a universe that exists. Therefore stability must dominate.

The electron model introduces the need for spin and magnetic considerations, since a spinning particle should create a magnetic pull through interaction with the storm. Storm god-photon/particle reactions create both spin and magnetism. Combinations of electrons or violin string particles both create a similar projection.

This projection has directionality, spin, and—in combination—both positive and negative magnetic anodes.

A simple design suits both projections—an arrow for directionality, a cross arrow for spin, and a letter designation to differentiate different particles.

We know from the standard universe how protons and neutrons are configured, so we assume a similarity to the PS Universe. The simplest combination models using electron blocks and the violin string model both yield protons and neutrons of amazing stability. When directionalities are aligned to a central point, the photon storm uses all its power to keep these elemental particles together.

The electron is drawn by its magnetism and directionality to the positive anode of the proton. Strangely, in the PS universe, the electron does not orbit the proton. Far from an orbit, the electron dances over the surface of the proton, with its negative charge semi-attached and its positive backside flying somewhat free. Again, the storm twisting around the electron causes both a negative and a positive charge in opposite directions.

If the projected design is at all accurate, then it explains why ions and atoms acts as they do, seeking new attachments. And in the helium configuration, the design explains why the atom is relatively non-reactive, since the attachments are as tight as they can get.

A question to ask is why the electron does not bind to both sides of the proton, both the positive and negative sides? They answer may be that the electron prefers not to bend as severely as these attachments would require; it would rather stretch than bend? Maybe the electron, like many humans, has a bad back from all its exertions. So the electron reaches out to other anodes further in distance but more comfortable to its configuration.

A twisting magnetic field must flow through every part of the configuration and it should generally balance into a single flow. In addition, the power of the photon storm should play a large role in keeping particles together, and in particular, pushing them into roundness. The roundness concept may explain a number of otherwise anomalies in the construction of atoms.

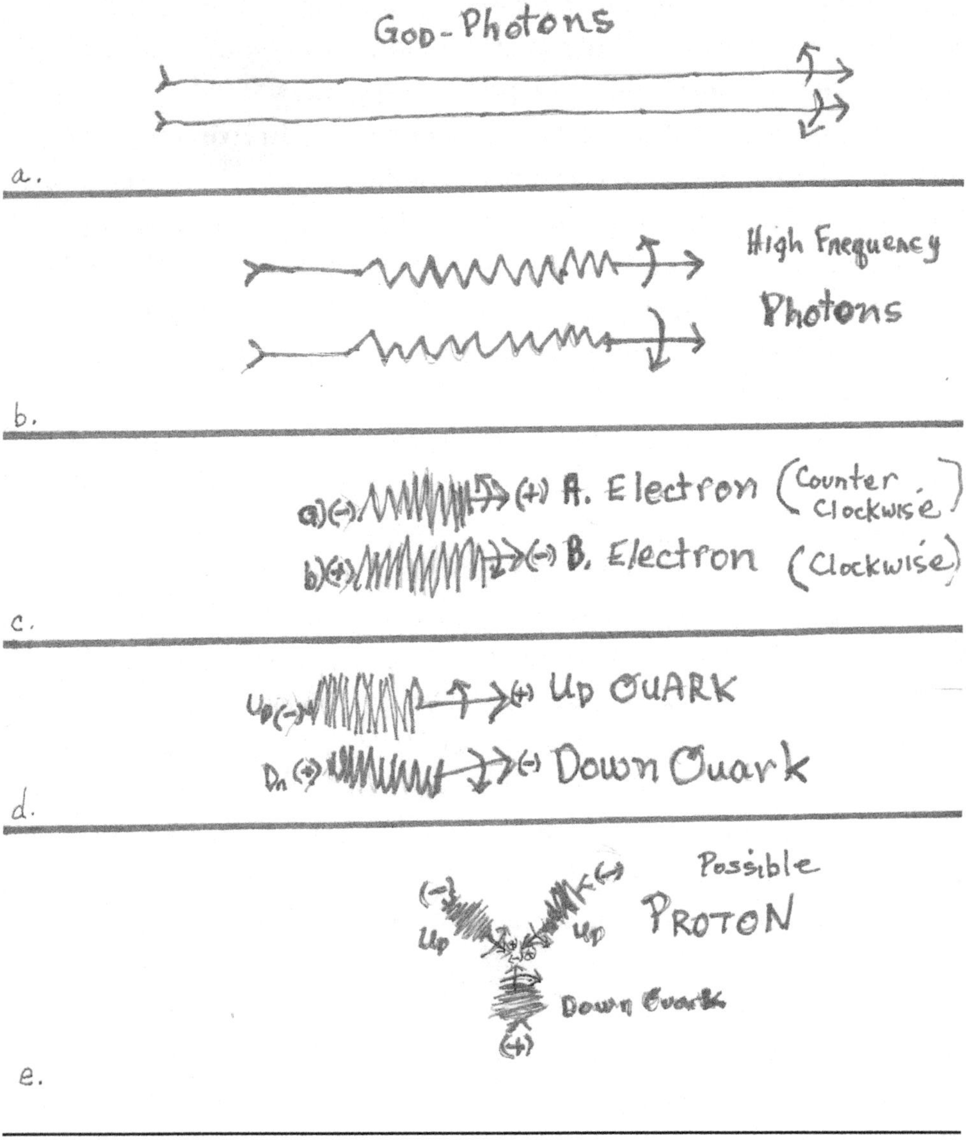

Goo-Photons

a.

High Frequency
Photons

b.

a)(-) /WWWW⇒ (+) A. Electron (Counter Clockwise)
b)(+) /WWWW⇒ (-) B. Electron (Clockwise)

c.

U₀(-) /WWWW ↑ ⇒ (+) Up QUARK
Dₙ (+) /WWWW ↓ ⇒ (-) Down Quark

d.

(-) K (-) Possible
Up (-) Up PROTON
Down Quark
(+)

e.

In the previous drawings, a-e, note the continuity of structure from photons to particles necessary in the PS universe--each particle has spin (a); increased energy or compounding (b) creates interference with the storm and acquires mass(c); particle spin creates electromagnetism by twisting the storm. Electrons spinning in opposite direction acquire opposite charge in line with their directionality. Finally (e) when elemental particles meet at the same point directionally, they unite with extreme power, continuing to twist around the same tri-point.

Note (e) also, an invisible magnetic current of twisted storm particles extends in both directions from the positive and negative nodes. (Not drawn)

CHAPTER 15 – THE PROTON, NEUTRON, ELECTRON COMBO IN THE PS UNIVERSE, CONSERVING MAGNETIC FLOW. HYDROGEN THROUGH NEON.

On the following page, the drawing is a schematic of a simplified Photon Storm proton (f), and a PS neutron (g). Note both particles entail a magnetic stream of photons moving circularly in both directions from node to node. (Not pictured) Also, these particles would be attended by fluff, photon semi-particles attracted and produced by their magnetic configurations. Fluff would provide 99% of the size and mass of the particles and if we could actually look at these particles, fluff is all we could see.

The reader should be advised, in the following drawings, accuracy is not really possible. Please examine the drawings for the sake of contemplation only and with the idea these may have *some vague connection* to a predictive reality.

These are rough sketches only. Please do not assume we actually know what we are doing. These drawings are inaccurate in many obvious ways. Hopefully, the reader can do a better job putting the parts together.

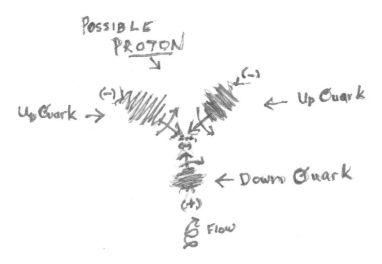

POSSIBLE
PROTON

(-) Up Quark

Up Quark →

(-) ← Up Quark

← Down Quark

(+)

Flow

f.

POSSIBLE
Neutron →

(+)

(+) ← Down Quarks

(-) (-)

(+)

← Up Quark

(-)

g.

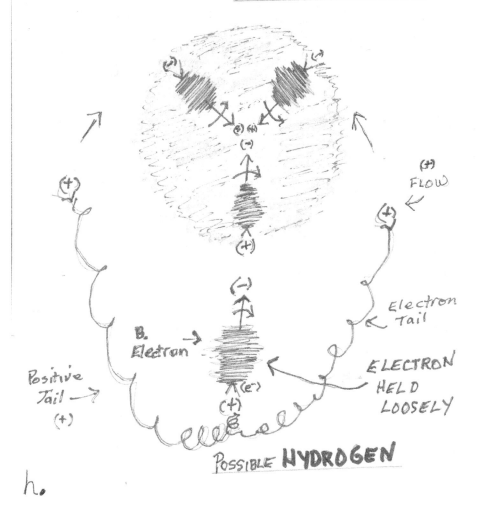

PROTON with ELECTRON

POSSIBLE HYDROGEN

h.

In the previous picture (h), a single electron attaches to the positive anode of a proton. The electron is drawn to the positive anode by its negative directionality. The positive tail of the electron is free to seek out a new attachment, either bending upwards towards the negative anodes of the proton, or stretching to the negative portal of a different atom.

The following picture (i) is a conception of a possible proton (dark), neutron (white) combination. Note this combination is quite strong with two positive and two negative nodes closely aligned. An electron is attached to the positive anode of the proton, with its positive tail seeking flow to a negative anode, either on the attached neutron or on another atom.

Fluff would flow freely through the tiny atom, actually a miniscule dot covered with Bourne-fluff.

(i)

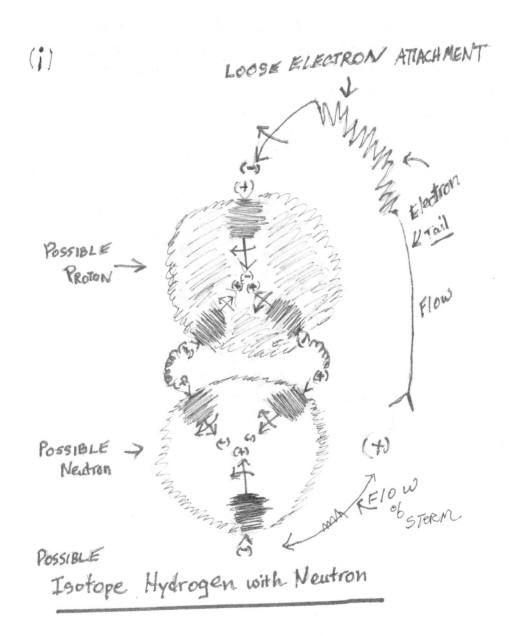

LOOSE ELECTRON ATTACHMENT

Electron Tail

Flow

POSSIBLE
PROTON

POSSIBLE
Neutron

(+)

FLOW of STERM

POSSIBLE
Isotope Hydrogen with Neutron

CHAPTER 16 – CREATING RATIONAL ATOMS AND MOLECULES

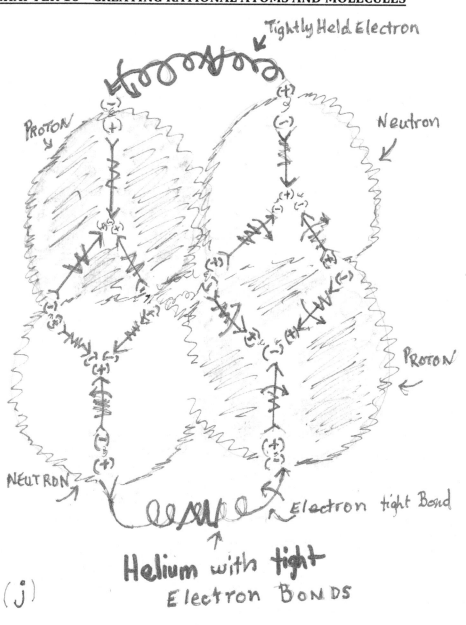

Helium with **tight** Electron BONDS

(j)

Picture (j) on the previous page is the possible construction of a helium atom in the PS universe, consisting of two (dark) protons, two (white) neutrons, and two electrons. Note the tight electron bond between anodes that would pull the atom together. Note also the complete storm flow throughout the atom united through the electrons in which twisted storm energy would circulate completely through the atom. Also, recognize, due to the tight electron bonds, that the helium atom would be quite reticent to join with other atoms, though a He2 would be quite tight also. Finally, note the completed atom would be pushed into a spheroid shape by the force of the storm from all sides and the magnetism pulling it together. The central atom would actually be tiny, tiny, tiny, the size of a single planet surrounded by a vast solar system of circulating fluff energy and spinning electron flares. The Technicolor version should be in theaters soon.

The following picture (k), is a proton-neutron combo as viewed from top, bottom, and side. Picture (m) is a possible lithium atom in the PS universe, basically a helium atom with an attached proton-neutron combination. The configuration has a fairly free electron attachment, so it would be looking to join with another atom. However, the configuration has a problem--it is very out of round. The helium center would be ball-like and the proton-neutron combo would not hang on very effectively. Most important, the non-spherical configuration would be intolerable to the storm of photons pelting it in every direction and struggling to push it into a sphere.

A more spherical configuration of lithium is drawing (q). Such a spherical lithium configuration—caused by the photon storm--might explain lithium's weight of 7, rather than 6.

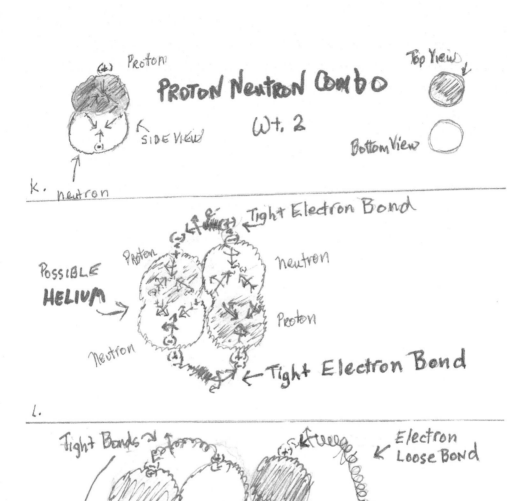

Proton

(+) Proton

SIDE VIEW

PROTON NEUTRON COMBO

Wt. 2

Top View

Bottom View

(-)

k. neutron

Tight Electron Bond

Proton

neutron

Proton

POSSIBLE
HELIUM

Neutron

←Tight Electron Bond

l.

Tight Bands

Electron
Loose Bond

(-)

(+)

(-)

m.

POSSIBLE LITHIUM w/ Loosely held Electron

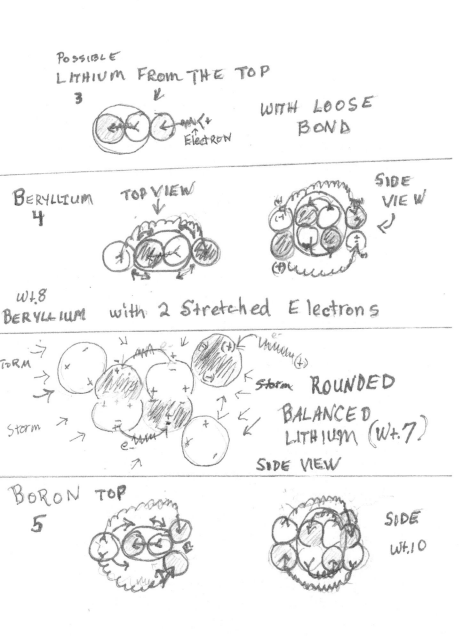

POSSIBLE
LITHIUM FROM THE TOP
3   ↓   WITH LOOSE
          ← ✕ ← ← +   BOND
                ↑
              Electron

o.

BERYLLIUM   TOP VIEW   SIDE
4   ↓   VIEW
                                              ↩
         (+)
        ← ✕ ← ← ✕ ←
          →   →

p.

wt.8
BERYLLIUM   with 2 Stretched Electrons

STORM →   e⁻   Uranium (+)
     →   ← Storm   ROUNDED
STORM →   BALANCED
q.   LITHIUM (wt.7)
   storm →
                e⁻   SIDE VIEW

BORON TOP   SIDE
5   wt.10
r.

In the previous pictures, note beryllium (p) has 2 obviously stretched electrons, best viewed from the side view and two obvious electron orbitals, both containing 2 electrons. A possible picture of spherical lithium, wt. 7, is pictured in (q). And boron is pictured in (r). Boron has a similar problem to lithium in not being very spherical. The storm might add neutron mass to boron's configuration to correct this fault, giving it a wt. of nearly 11, or the same as it possesses in the classic universe.

ORBITALS: *The PS model does an excellent job creating the s orbitals, up to four electrons in two roundish orbitals. The p orbitals are more problematic. Why aren't they round? Well, the added proton-neutron combos are all in the same circle around the helium core until neon. The orbitals must compete for space. In neon, the core is surrounded completely, and a new shell circle must begin at the next element with a proton-neutron* <u>outside</u> *the neon core.*
Such a <u>physical</u> *model might be a nice teaching guide?*

In the following pictures, a possible carbon atom (s) configures reasonably with four obvious semi-loose electrons. These electrons explain carbon's ability to combine as it does.

The nitrogen atom (t) is more difficult to figure out since where to put another proton-neutron pair is not obvious. If a pair is placed where pictured, it might replace one loose bond with a tighter one. Thus three electrons remain semi-loose.

The oxygen atom (u.) might configure in several ways (two are drawn). But both configurations have two very loose bonds. Either configuration, if accurate, would explain oxygen's natural tendency to join with other atoms like hydrogen in the PS universe.

POSSIBLE CARBON

6

Wt. 12

S.

TOP

4 Semi-loose Electrons

NITROGEN

7

t.

3 Semi-Loose Electrons

Oxygen   or   TOP

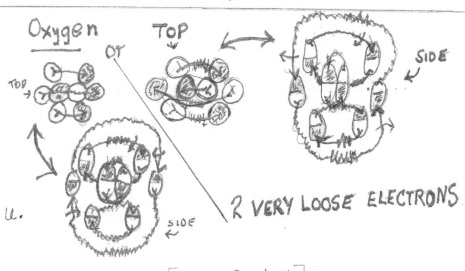

TOP

SIDE

SIDE

2 VERY LOOSE ELECTRONS

u.

2 Configurations   [Oxygen 8 - Wt. 16]

The following drawings are the possible atoms of fluorine and neon. Fluorine has one very loose bond, accounting for the way it connects to other atoms.

The simplest configuration for neon, drawing (w) has only the shortest electron bonds. The original helium core is completely surrounded as shown by the top view provided. In fact, five very stable helium/alpha particles make up the entire atom. Thus the reason neon does not require other atomic unions should be absolutely clear from the model.

We could design further but we are running out of crayons. *Hopefully the point is made that the photon-storm criteria generates particles with predictable properties. And these particles share a confluence with the classic universe, which may be useful.*

We must repeat again that these drawings are not particularly accurate. They are sketches; the way atoms actually pack together surely may vary. However, regarding electron orbitals—orbitals are totally dependent on how proton-neutron pairs pack together.

Until we understand atom packing, we can't really understand orbitals. And to study atom packing, we need to assume protons pack together in some reasonable fashion.

And finally, the mathematics of fluff might be productive, since fluff is most of the PS universe. Fluff packs on particles like ice on snow or mineral crystals. Each configuration allows a slightly different variation. It may be that every single atom in the universe, like snowflakes, might be a bit different?

FLUORINE 9

Freeish →
Electron
on
Bottom

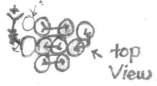

↖ top
View

V.

Wt. 19

POSSIBLE Neutron extra

SIDE
VIEW

1 very loose e⁻ bond

NEON 10

No Free
Electrons

W.

Wt 20

CHAPTER 17– SOME MORE SPECULATIONS, INFINITIES AND BEYOND, AND HOW TO TEST THE CONCEPT

The Photon-Storm Universe remains speculative. But we feel the argument has been made for many possibilities of entanglement with our classic universe. The reader may decide whether the similarities are coincidental or merit further analysis.

And some questions can be addressed in the Photon-Storm universe with greater clarity than in the standard universe. Here are some:

WHAT DOES A PS UNIVERSE LOOK LIKE?

The PS universe, as opposed to the standard universe, has an edge surrounding it. An edge of some sort is necessary to maintain the extraordinary amount of energy in the PS universe.

A vacuum may actually form the edge of the universe. Perhaps photons, dependent on conjoined spin to move forward, cannot move through a vacuum. [See pgs 25, 26 – the Spin Theory]

And if photons cannot stay together in a vacuum, matter would have an equally impossible time. In the PS universe, energy is necessary to keep matter together. And all other forces including gravity and magnetism also depend on the presence of the storm.

INFINITY AND BEYOND, IN THE PHOTON STORM UNIVERSE

Infinities are difficult problems for human brains, since we cannot mentally imagine something infinitely large or small. In our classic universe, space must be infinite. In the familiar universe, empty space goes on forever, because, well, what else can empty space do? However, in the PS universe, space infinity *may not* exist.

The may *not* exist scenario is dependent on the true concept of nothingness. If a true vacuum does not exist in the PS universe, then space is defined completely by the photon storm. Thus, when not defined by the storm, vacuum-space simply does not exist. Nothing is actually nothing. It isn't there.

If we imagine a defined edge around the entire PS multi-universe complex, then at this edge all God-photons turn around and return along the edge or inward to inside the universe. Thus, outside the universe exists nothing, not even space.

This nothing is absolute and profound. If we attempt to stick a measuring tool through the PS edge, the stick will dissolve at the edge. Matter cannot exist outside the storm in the PS universe. The storm is what keeps matter together. In like fashion, we cannot use a lazar device to measure nothingness. Light cannot move without the assistance of fellow photons inside the PS universe.

After trying *all* technologies, we can simply assume, nothing does not exist. Nothing is, in fact, truly nothing. Nothing—unlike a vacuum-- occupies *no space.* Thus the size of the PS universe is defined only internally and nothing externally exists. Outside is truly nothing. So, the infinity of space is nothing to worry about.

However, the edge of the PS universe *may* not be smoothly defined. If so, though the math is more complex, the result is similar. Somewhere, outside the PS multi-universe, nothing exists—or rather, nothing *doesn't* exist because it is nothing.

We can complicate the issue further, though, if we imagine ourselves standing outside the PS universe in the nothing that does not exist. We know we can not exist there, and this gives us pause, but still, if we look at the PS exterior, strangely we see exactly what we would see if we are looking at a black hole. In other words, we see more nothing, total blackness. Nothing comes out. So, the natural question to ask is, is a PS universe actually some sort of black hole? And further— stop me please—are all black holes capable of defining space in different fashions? Indeed, can a theoretical black hole define within itself a universe or even multi-universes?

The easiest answer is no, black holes are profoundly different than the PS multi-universe. The PS multi-universe is not a black hole.

But as Plato says, "Who really knows?" If black holes can somehow twist space through invisible wormholes, they may send space-defining energy via multi-dimensions absolutely anywhere, so that outsides become insides. In which case the infinity problem might grow more complex, but at the same time, the universe itself would grow even more interesting and exciting.

And by the way, the universe seems fairly exciting already.

THE PENULTIMATE PARTICLE

A speculation: For the sake of clarity, we have envisioned the god-photon to be the penultimate particle. Penultimate in our view means that the god-photon is the sole original particle; nothing smaller exists in this scenario; so everything builds from the god-photon.

The possibility exists, of course, that the god-photon as so imagined is not the penultimate particle. For example, a single god-photon can be a configuration of countless mini-god-photons, perhaps trillions of them. And if god-photons have a stable configuration far above gamma rays, then a mega-god-photon may exist also. And further, an entire multi-verse composed of not god-photons, but maga-god-photons might exist. Perhaps such a high-energy multi-verse is even defined within a black hole? Did we mention, this is idle speculation?

*We actually have an alternate, simple configuration for black holes. Stay tuned.

Picture following page: At the edge of a true vacuum, god-photons must turn along the edge or back into the universe.

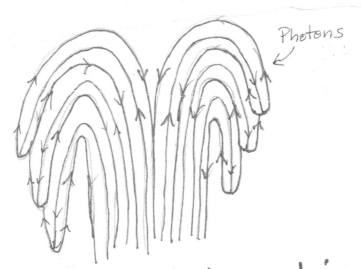

Vacuum

Photons

Photons Spin back on reaching
vacuum, pushing off each
other to reverse.

But—before we reach infinity--a vacuum is only one alternative--the
edge of a single PS universe may be formed by meetings with the edges of different

universes of wide probabilities, including something like our standard universe at a different energy level.

Alternate universes and non-universe constructions may exchange energy to explain the photon storm. In a grand speculation, the edge of the PS universe may be the inside of a black hole, and the photon storm may be the remains of what this dark hole has consumed.

We should stop before we hurt ourselves, but the reader should feel free to speculate on such things. They may find the answer.

EXPANSION AND DIMINUTION IN THE PHOTON STORM UNIVERSE

NOTE:

If the center of the universe is far denser in energy than the outskirts—which would happen if photons are bouncing indiscriminately from the edge—such a central pressure would push all matter outwards with increasing acceleration. This phenomenon, strangely, is what scientists have measured in our standard universe.

NOTE ALSO:

Whether the photon storm universe in total is expanding or growing smaller has very little to do with the amount of matter it contains, as many in the classic universe feel is the case in their universe. Expansion or not is totally controlled by the movement of the photon storm. The amount of mass has little to do with it.

Expansion of the matter in the universe may be assumed if more energy concentrates in the middle of the universe than at its edges. This expansion may occur even if the universe in total is not expanding.

However, the accumulation of energy at a central point could easily project a black hole of inconceivable proportions, or an amalgamation of such things. This central spot or spots could well be sucking in the storm faster than it is being extruded, and the final effect, if the storm disappears—taking with it all mass and gravity--would be a rapid re-expansion of a great gob of energy ready to recreate mass and gravity and pushing a new constellation of universes to new creations or lovely repetitions.

This all depends on how God wishes to play her music.

And is the speed of light stable, or is it changing imperceptibly in some areas? Is the total energy diminishing or growing stronger? What would this tell us?

And how would we ever know locally, since every possible measuring instrument is tied to light speed?

And also…?

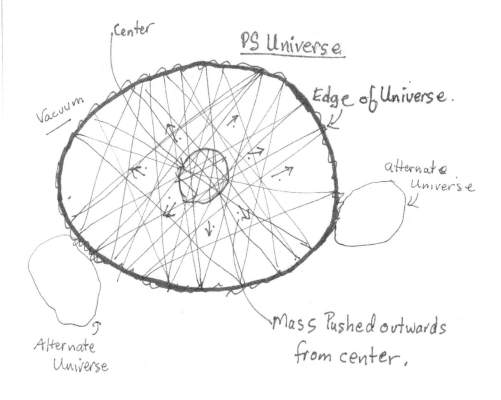

Center

PS Universe

Edge of Universe.

Vacuum

alternate
Universe

Alternate
Universe

Mass Pushed outwards
from center.

STORMS IN THE PHOTON STORM UNIVERSE

If we view the PS universe as a huge cloud or groups of clouds, then the photons in each area of the cloud move at the speed of light, stimulated to do so by the energy that surrounds them. This energy may differ from cloud to cloud, and thus so would the speed of light.

The entire cloud, however, could also be moving at *any* imaginable speed. And different parts of the cloud could be moving or swirling in different directions, all according to forces not yet understood.

Thus, though light speed is difficult to surmount within a given cloud, choosing a different storm wave might allow future, future generations to surf around the cosmos even faster than the speed of light.

Silver surfer? Well, why not?

And if tiny spinning electrons can create magnetic tornadoes, shouldn't spinning galaxies and entire universes also stir the cloud into ginormous magnetic torrents, torrents that push or pull entire universes into interactions? And?

LIGHT SPEED

And in the photon storm universe, the speed of light is not magical. If the density of the storm varies, or the tension on the v-string varies, or if God had struck the chime of the universe more erratically, the tension might be even higher, and so would the speed of light. Such speed only represents the shared tension of the entire universe divided by the number of violins.

So, if light speed has no special priority, it could be different in different parts of the universe—or it could be spread unevenly in distant bubbles. Such a disparity might allow future generations to increase or decrease light speed to specifications. If light speed could be decreased to zero or even switched to the minus category, time travel could be common...or maybe not...or...well...um...

BLACK MATTER AND BLACK ENERGY AND BLACK HOLES

In the standard universe, black matter and black energy are necessary to explain the amount of gravity, which needs to be present to explain measured phenomena in galaxies, namely that stars on the outskirts of galaxies move much faster than they should. To move as fast as their measured speeds, about 6 times as much matter must exist to provide enough gravity to keep stars from flying away at the outskirts of galaxies.

But in the Photon-Storm universe mega-amounts of energy circulate everywhere, much more energy by many degrees than predicted in the standard universe.

And gravity is supplied by the photon-storm in the PS universe; gravity is not simply a product of matter. Gravity proceeds from outside to in, rather than inside to out in the Photon Storm universe.

BLACK HOLES, incidentally, would not be totally black in the PS universe. Photons moving inward toward a black hole actually facilitate outward movement away from a black hole, so energy could leave a black hole—assuming the incoming energy density is not so great as to limit this phenomenon. In addition, matter that absorbs energy also continues to emit energy, so a black hole would emit energy particularly at its surface in the PS universe.

THE INCREASE OF GRAVITY AT THE EDGE OF GALAXIES.

Presence of the Photon-Storm might also explain gravity anomalies noted in galaxies. The edges of galaxies speed much faster than predicted, assuming gravity is less at the edge of a galaxy than in its core. For more than 100 years, scientists have questioned this observation. But in a photon-storm universe, the storm is actually the source of gravity, not mass. The computations are in line.

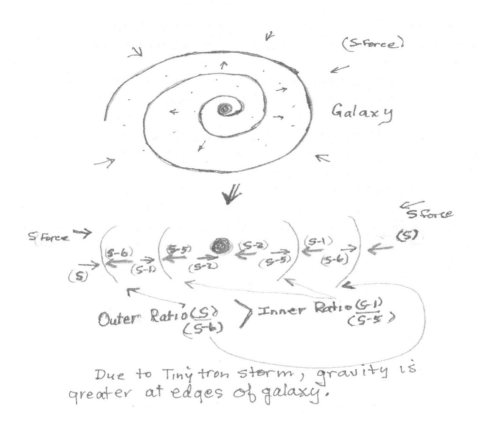

Gravity ratio is greater at the galaxy edges, and the requirement for dark matter is diminished.

<u>STAR POWER in the PHOTON-STORM UNIVERSE</u>

And a number of questions still exist concerning stars. When, for example, we consider a star in our universe, the star's power is overwhelming—a huge ball of energy pushing protons together to create larger elements, expelling massive amounts of energy in the process.

Stars *appear* much the same in the Photon-Storm universe, but their awesome star power is dwarfed totally by the far greater force of the invisible photon-storm.

In the PS universe, a star's shape is caused by the photon-storm pushing it into roundness. In addition, the star's gravity, the reason its elemental particles have come together, is also due to actions of the photon-storm.

And most importantly, the energy a star expels is actually provided initially and constantly by the photon-storm. Mathematicians in photon-storm universes should be constantly perplexed by what we consider star economics in our standard universe. If e=mc2, how can any star continue to expel massive amounts of energy and ions without losing their mass?

In the PS universe, stars have an easier time and longer lives, because their reserves of energy are constantly replenished by photons from the storm.

High-energy protons and neutrons absorb extra energy from the storm; when they squeeze together to make helium, the energy is released as high-energy photons.

Indeed, we may view a star in the PS universe as an energy converter, rather than an energy producer. A star converts the energy of small photons into the energy of large photons, while producing matter as a by-product.

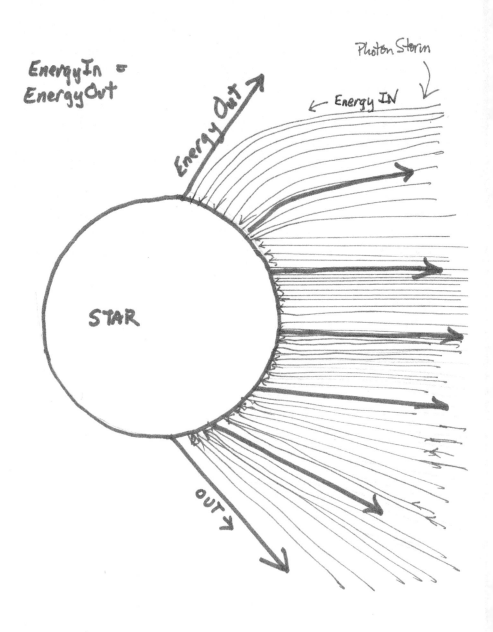

Energy In = Energy Out

Energy Out

Photon Storm

← Energy IN

STAR

OUT

Energy in from the storm is converted to high-energy photons going out.

STAR SURFACE TEMPERATURE

Also suggested by the previous drawing, much of the activity of a star in the PS universe takes place near the star's surface. The storm has been squeezed together intensely, accelerating particles toward the star at near light speeds, intercepting high intensity photons and ions that push outward from the star. If the pressure is not as high as in the star's core, it is still high enough to press hadrons together, creating elements and perhaps even creating original matter.

The several processes, including twisting photon tornadoes, all create a tremendous amount of energy. That the coronas of stars are hotter than their interiors seems very reasonable with this model.

GROWING OLDER

As elements in a star grow larger, they grow less and less affective as energy converters. A proton's participation in a large nucleus makes its surface area to interior area ratio increasingly smaller in size and therefore less able to absorb or emit energy at its surface. This situation increases as the number of protons in a nucleus increases according to the *liquid drop rule*.

As a star matures, it loses its ability to act as an energy converter. Its fuel of small elements becomes exhausted. And larger nucleuses when combined do not expel energy as robustly. If fact, beginning with iron, elements drain the star of energy when they are being created.

STACKING BLOCKS

The structural mechanisms by which large nucleuses grow unstable are somewhat vague. Spin is the weak force in the PS universe that tears apart atoms, and all parts of an atom tend to spin. When children stack blocks, they quickly realized the instability of their pile grows as the height of the blocks rises higher.

Block stability is a function of: #blocks, Height of the pile, width of blocks, and K, (a constant containing room breezes, block design, destructive siblings, etc.)

$$\text{Stability} \sim \frac{width}{ht}(k) \cong (k)width/\#blocks$$

So stability decreases with the number of blocks on the pile.

The same sentiment can guide us in assembling nucleuses from atomic blocks in the simple Photon-Storm universe. Remember, physics is simpler there.

Height and width are probably not vectors in atomic nucleuses, but a *similar* phenomenon possibly occurs.

Stability $\sim \dfrac{w}{h}(A3)(k)$

Stability is a function of :

w=ratio of the surface area of a sphere@radius/volume of sphere

@radius= $\dfrac{4\pi r2}{4}/3\pi r3$

h=#hadrons

(k) = constant related to hadron stacking mechanisms.

#of Spins: OR STATED SIMPLY, STABILITY DECREASES WITH THE SPIN VARIABILITY OR NUMBER OF SPINS.

As r increases w/h grows smaller since the volume of a sphere increases with the cube of r and the surface area increases with the square of r.

So nucleus stability generally decreases with increased size due to spin variances.

Using a visual model, the photon-storm pushes a spherical element together at its surface. As the element grows larger, pressure from photon spin crosscurrents increase compared to photon-storm pressure directed specifically at the center of the sphere.

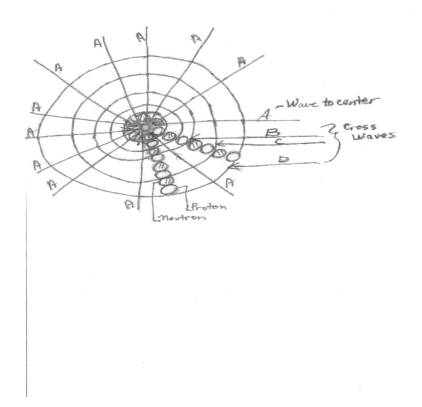

Assuming hadrons keep the integrity of their separate and round shapes, then a large nucleus may act like a bag of spinning balls. Like model trains running around the same tracks--the more trains, the more opportunity for a collision which could break apart an atom. Assuming a collision event could happen randomly at ten year intervals, then the atom would have a half life of ten years.

Increasing freedom should allow hadrons to expel Bourne-fluff energy absorbed earlier from the storm during atomic construction, so if particles break apart through a fission event—an event in which the particle spins apart--and they resume smaller compact states, this extra energy can be expelled.

WHAT IS THE VALUE OF S?

S is equal to the force of the photon storm over a specific area. Such a value is important in the PS universe to determine mass and gravity. Unfortunately, the scientists who dwell in the PS universe have not communicated their computations to our universe.

However, if we jump to the assumption that the PS universe and the standard universe are similar, then we could use a number of useful measurements physicists have arduously revealed over the years.

The proton's mass has been calculated to a high degree, as has the size of the proton to some accuracy. Since, in the PS universe, (S) (Rk) (A3) = mass. Then (S) = mass of electron/(A3)(Rk). If we assume (Rk), the reflectivity of the proton particle, is 100%, or a value of 1, then (S) = mass of proton/(A3 or size of proton) (1). (S) = mass of electron/(A3).

Or, more simply, the value of (S) would equal the measured mass of a proton divided by its measure size. The error in this method might be large, however, since reflectivity might not be 1, and both mass and area values are questionable. At best, such a method would give an estimate of the great power of (S).

Bear in mind: the added fluff might be of a different reflectivity than the quark combinations. Such a situation might well explain energy anomalies measured in accelerators.

Utilizing a larger particle, such as a planet like earth, might be easier for a very rough estimate of (S). The mass and area are roughly measured.

Assuming (Rk), the standard reflectivity is equal to (4)(10)-13, the empty space in a single atom, the (S) = Mass of Earth divided by the area of earth and the guessed value of (Rf). (S) = (5.9722)(10)24Kg divided by [(10)9 cubic meters, the cubic area of earth, times (4)(10)-13, the empty space.

Then:

(S) =(5.9722/4)(10)28nd Kg/cu.m or (1.493)(10)28 Kg/cubic meter
This is quite a strong force.

CHAPTER 18 -- A SIMPLE TEST FOR A SIMPLE UNIVERSE

Discussions of the PS universe often reach the same conclusion: The PS universe is an interesting concept, but nearly impossible to prove. That the Standard Universe and the Photon-Storm Universe share almost all mathematical symmetries only makes the quandary more difficult.

However, some unshared qualities might be measureable. In the PS universe a magnetic field occurs because photons are bounced in a certain direction. Though only a tiny portion of the storm is bent, the effect is magnified by the tendency of God-photons to bend together.

Performing the double-slit experiment might offer some validation of a PS universe, since the affects are easily explained in that universe. (See Pg. 5)

Also, if the PS model is correct, then an object blocked by a cylindrical magnetic field will receive slightly fewer collisions at its sides. A non-magnetic sphere in the middle of such a field should grow wider in the PS Universe and the result might be measureable.

But the magnet should have no effect in the Standard Universe—if that is our universe.

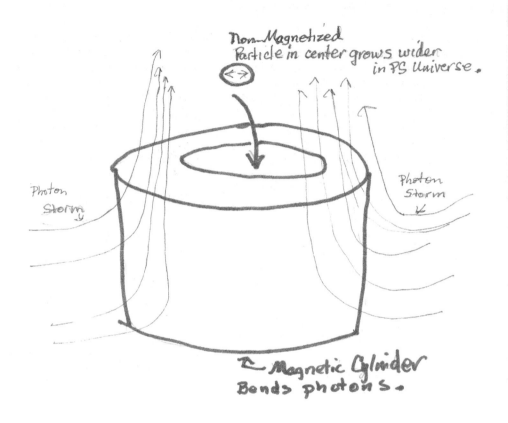

When the storm is blocked from the sides, an object may grow wider in the PS universe.

Also review the double slit experiment in chapter 1. The experiment actually makes sense in the PS universe.

MATHEMATICAL MODELS

Deriving mathematical models to explain phenomena in the PS universe is many times easier than in the classical universe. The PS model is easy to visualize and thus easier to re-model mathematically. Suffice to say, the mathematical formulas in the PS universe may not be unlike those in the classical universe. Later works may have more time to explore the duality.

SOME PRACTICAL APPLICATIONS

Often, when preparing this argument, any practical application seemed unlikely---at least not in the foreseeable future. We are already utilizing the photon storm affects in extraordinary ways, no matter the model of our understanding.

However, if we recognize the incredible power of the photon-storm moving through each centimeter of space, if only a tiny portion of that energy could be realized, the human race would be greatly rewarded—assuming it does not recognize this energy as another way to destroy itself.

A photon storm—if it exists—is intimately involved in every material and chemical process from the tiniest atom to the largest galaxy. Understanding this process might be more easily taught in a PS universe model and might lead to helpful innovations not now considered.

MATTER CREATION

A most interesting phenomenon in the PS universe is the creation of matter. The consensus in the classic universe is that an immense amount of energy would be necessary to create matter, the energy of a huge bang.

However, if the point stability concept suggested by the PS model is correct, creating matter might not require a great deal of energy at all. Rather, matter creation is a game of finesse. If tiny photons are directed at the appropriate point, matter should appear.

Granted, this point is infinitely tiny and the technological needs would be significant, but in theory matter creation is possible using less energy than the created matter might contain. $E \leq mc2$ The extra mass/energy might be donated free of charge by the photon-storm.

ENERGY PRODUCTION

And also in theory, the finesse forces utilized to create matter could also be utilized to create clean energy. At near-light speeds the force holding atoms together diminishes significantly. And if anti-matter can be easily created or released from normal matter, the ability to create energy would be nearly endless. Note: In the PS universe, the only difference between matter and anti-matter is spin, and spin only.

And a magnet drive might actually work. A spinning mechanism could create a strong magnetic field that reacts like a single photon in only one direction against the storm, and in certain low-friction situations, it could pull or push a ship or tram pushing against the storm in the same manner as a photon, with very little energy use. And then, we all remember those flying saucers.

And...

Thank you very much for your patience.

IN CONCLUSION

The purpose of this essay is to promote a discussion of a Photon-Storm universe. We hope enough has been established to encourage interest in the model.

The premise may be helpful in part even if questions remain. And if inaccuracies exist—and certainly they do—then everyone is invited to consider what corrections would be helpful.

The advantage of a simple model is that everyone can play with it.

Thank you,

Comments appreciated:

Ralph Bourne
ralphbourne@sbcglobal.net
78 N. Corona Dr.
Porterville, CA. 93257
559 920-1976

Made in the USA
Las Vegas, NV
28 December 2020